THE
WELDER'S
BIBLE

Acknowledgments

No book can possibly be the work of a single person. My book about welding is certainly no exception. I would like to thank in particular Glenn Spoerl of Sears Roebuck & Company for the equipment used in many of the illustrations. I would also like to thank Airco Welding Products, Hobart Brothers Company, Lincoln Electric, Modern Engineering Company, Charles R. Self and Victor Equipment Company for supplying me with technical information and illustrations.

I would also like to thank Peter Monaco for help in shooting the photographs that I included in the book. Last, but certainly not least, I would like to thank my wife, Virginia, for encouragement in times of need and for help with the final typing.

Dedication

To all do-it-yourselfers who are looking to expand their range of abilities through doing and learning.

THE WELDER'S BIBLE

BY DON GEARY

TAB BOOKS Inc.

BLUE RIDGE SUMMIT, PA. 17214

FIRST EDITION

FIRST PRINTING—JULY 1980
SECOND PRINTING—DECEMBER 1980
THIRD PRINTING—MAY 1981

Library of Congress Cataloging in Publication Data

Geary, Don.
    The welder's bible.

    Includes index.
    1. Oxyacetylene welding and cutting—Amateurs' manuals.
II.
Title.
TS228.G42   671.5'2   80-14562
ISBN 0-8306-9938-4
ISBN 0-8306-1244-0 (pbk.)

Cover Courtesy of Hobart Brothers Co.

Introduction

In this book you will learn the basics of metal joining and cutting using the oxyacetylene process. I should mention, at this point, that I do not advise the do-it-yourselfer to teach himself how to weld with any method other than oxyacetylene. While the text is designed for the do-it-yourself welder, there are many references and explanations as to how welding and cutting is done by professionals — men who work at metal joining everyday. I trust you will find these additions helpful in learning how to weld. If any terms are unfamiliar to you, check the glossary. I have tried to make it as complete and helpful as possible.

I will caution you in the beginning, as well as throughout the book, that working with oxygen/acetylene welding or cutting equipment can be dangerous. You, as a welder, must always approach any welding problem soberly. Use your best judgment at all times. Safe welding and cutting should always be your first priority. In this way you will develop sound and safe welding habits that can only result in a long and healthy welding experience. Safety is of such a concern to me that I have devoted an entire chapter to pointing out how to safely handle welding equipment, as well as how to deal with mishaps such as fire around your welding shop. Let me suggest that you read this chapter before setting up your welding equipment or lighting your torch for the first time.

While it is entirely possible to become a proficient welder solely with the aid of this book and many hours of practice, you should supplement the information contained in these pages with advice and instruction from a professional in the art of welding. Spend the time to seek out professional welders in your area. These men are often warehouses of both practical and technical information. If time permits, you should also enroll in some type of educational program that deals specifically with welding and cutting — adult educational programs, offered by continuing educational departments in your area, for example. Often the cost is minimal and you will get the opportunity to learn some of the things that cannot possibly be found in any book about welding.

If you do not at this time own an oxyacetylene welding or cutting torch, but are considering buying one, buy the best you can afford. I suggest that you pay a visit to a welding supply house in your area. Look in the yellow pages and ask for advice. More often than not, the people who work at supplying welders with tools and materials are very helpful to beginners. At the present time a basic welding and cutting outfit can be had for about $150. Used equipment, in good condition, may cost half as much. As far as I am concerned, the only way to determine if used welding equipment is in good condition is to have it checked over by a professional who specializes in welding equipment.

It will take you many hours to learn how to join or cut metal with the oxyacetylene process. Devote the time to learn how to do things properly. In the long run this is the best approach.

Now it is time to take a look at the oxyacetylene welding process. Safe welding.

Don Geary

Table of Contents

Chapter 1
Oxyacetylene Welding Fuels

Welding is a relatively new way of joining two or more pieces of metal in such a way as to make the finished piece as strong as the original metal. The oldest type of welding – *oxyacetylene welding* – was developed around the turn of the 20th century. There have been many developments in metal joining processes since then (Fig. 1-1).

For our purposes, metal joining can be broken down into three rather broad categories — *gas welding*, *electric welding* and *gas/electric welding*. Since the home welder is not interested in production work, there is little need, other than for informational purposes, to learn about other types of welding such as *laser* or *plasma welding*. These processes are used solely in industry (Fig. 1-2).

HISTORY OF GAS WELDING

Gas welding, the oldest and easiest of all types of welding to learn about metal joining, is very simple in principle. Basically, oxygen and acetylene are burned together to produce a flame which is hotter than the melting point of most metals. The temperature of an oxyacetylene flame is generally accepted at being around 6,000 degrees Fahrenheit.

In view of the fact that oxyacetylene is so widely used, it is almost unbelievable that this process did not come into existence until the beginning of this century. Oxyacetylene welding was first made possible through the experiments and discoveries of a French chemist, Le Chatelier, in 1895. He was the first to discover that burning oxygen and acetylene produced a flame with a temperature far higher than that of any other flame in existence.

It wasn't long before the capabilities of oxyacetylene became known to the industrial world and were put to use. After a workable way to store and transport oxygen and acetylene was developed, the road was clear for widespread use of this new method of joining metals.

1

Fig. 1-1. Modern welding in progress.

Fig. 1-2. Some forms of construction would be almost impossible without modern welding techniques and equipment.

It was probably the World War I that accelerated the use of oxyacetylene welding. The pressures to supply a fighting army and repairs to existing heavy equipment brought oxyacetylene welding in contact with millions of people. After the war, there arose a need for greater controls over the welding process and machines were developed that could weld (Fig. 1-3).

The oxyacetylene welding/cutting process is the most versatile means of working with metals. No other equipment or process in use by the metal industries is capable of performing such a wide variety of work on most types and thicknesses of metals. The oxyacetylene method of welding is also the easiest to master and probably the most versatile for the do-it-yourself welder. The bulk of this book will be concerned with this type of welding.

It is obvious from the name, oxyacetylene, that this type of welding uses a combination of oxygen and acetylene. From the standpoint of knowing about the basic components, it may be helpful to discuss these two substances and explain their significance to the welding process.

OXYGEN

Oxygen is present in the air we breathe but in small amounts. About one-fifth of our atmosphere is pure oxygen. Oxygen used in the welding process is about as pure as possible — over 99 percent pure. It is interesting to note that the process which results in pure oxygen for welding and medical purposes is called the *liquid-air process.*

Liquid-Air Process

To oversimplify the liquid-air process, atmospheric air, as mentioned earlier, consists of about 20 percent pure oxygen, 78 percent nitrogen and 2 percent other gases (by volume). Oxygen and nitrogen have different boiling temperatures. Thus, it is easy to separate the two by heating atmospheric air to a certain temperature and holding it at this temperature until the nitrogen, which has a boiling point of 295 degrees Fahrenheit, boils off. After the nitrogen has been removed from atmospheric air, oxygen and a small amount of other gases remain. These include carbon dioxide, argon, hydrogen, neon and helium. Since oxygen has the highest boiling point of all these gases, the remaining mixture is further heated until only pure oxygen remains. The pure oxygen is then stored as either a gas or liquid, depending on the eventual use. The liquid-air process is probably the most widely used method of producing pure oxygen.

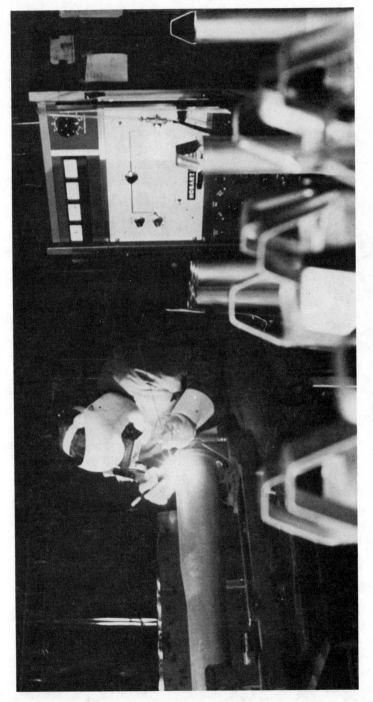

Fig. 1-3. Welding has many applications in industry today.

Fig. 1-4. The Spitfire torch, while not a true welding outfit, can be used for many light duty projects.

Solid-Ox System

As a side note, I should mention another method used to produce oxygen. This method requires the use of a special oxygen bearing pellet which is burned and gives off oxygen. The pellet is one of the main components of the *solid-ox* system and is sold in lumber yards and variety stores throughout the United States. While the solid-ox welding system is rather limited in capabilities, it has nevertheless made light welding and brazing available to thousands of do-it-yourselfers for just a few dollars (Fig. 1-4).

Cylinders

Oxygen is commonly sold in cylinders in three sizes: 244 cubic feet, 122 cubic feet and 80 cubic feet (Fig. 1-5). There are very strict requirements for oxygen cylinders. Each must be able to withstand over a ton of pressure per square inch. The Interstate Commerce Commission (ICC) has set up guidelines for oxygen cylinders. No part of the cylinder may be less than 1/4 inch thick. Each cylinder

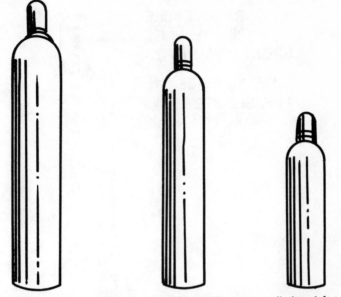

Fig. 1-5. The three most popular sizes of portable oxygen cylinders, left to right, are 244 cubic feet, 122 cubic feet and 80 cubic feet.

must be made or forged from a single piece of steel. The steel itself must be armor plate, high-carbon steel (Fig. 1-6).

Since the ICC requires periodic inspection of oxygen cylinders, which are shipping containers, very few oxygen cylinders are actually owned by private individuals. The oxygen supply houses, which own and lease oxygen cylinders, have the responsibility of complying with ICC regulations and inspections. This makes the whole affair easier on the consumer.

It has been my experience that you can lease oxygen (as well as acetylene) cylinders for a fixed period of time. When you require more oxygen, you simply bring the empty cylinder to the dealer and he replaces it with a full one. All you pay for is the oxygen. The oxygen itself is what you lease. In most parts of the country a 20-year lease is available and quite reasonable. In my case, the cost of a cylinder of oxygen and a cylinder of acetylene costs about $9 a year. Oxygen and acetylene are, of course, extra.

The universally accepted color for oxygen cylinders, lines and control knobs is green. Since there is no regulation which requires oxygen cylinders to be green in color, many companies paint their cylinders a special identifying color. It will be to your advantage to become familiar with the color used by your dealer. Often two oxygen supply houses in the same city will have two different colors for their tanks.

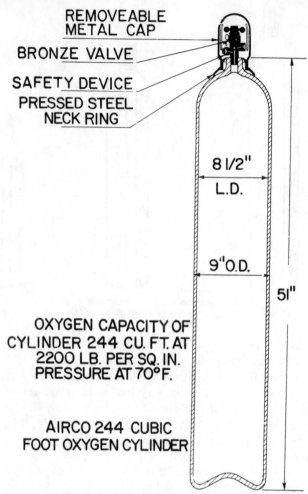

REMOVEABLE METAL CAP

BRONZE VALVE

SAFETY DEVICE

PRESSED STEEL NECK RING

8 1/2" L.D.

9" O.D.

51"

OXYGEN CAPACITY OF CYLINDER 244 CU. FT. AT 2200 LB. PER SQ. IN. PRESSURE AT 70°F.

AIRCO 244 CUBIC FOOT OXYGEN CYLINDER

Fig. 1-6. A typical oxygen cylinder.

Safety Practices

Oxygen cylinders are not dangerous when used and stored according to generally accepted safety precautions. Nevertheless, some general comments are in order.

Use only cylinders carrying ICC markings. You can be certain that cylinders of this type comply with the stringent regulations of the ICC.

Store cylinders only in a safe location, and fasten them in place. This will insure that the cylinder cannot be knocked over. Keep tanks in an area away from stoves, radiators, furnaces or other overly warm places. Oxygen cylinders should also be kept away from all

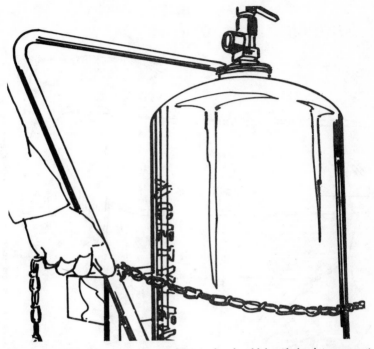

Fig. 1-7. Both oxygen and acetylene tanks should be chained to prevent accidental falling.

combustible materials or liquids. If cylinders are stored in the open, they should be protected from the elements of water, heat, cold and the sun's direct rays.

Oxygen cylinders should be stored and used in an upright position. In most cases a hand truck, specifically designed to hold two cylinders, is the best means for storing and using oxygen. The cylinders should be chained or strapped to the cart to prevent the cylinder from ever falling over (Fig. 1-7).

Never use the valve on top of the cylinder to lift the cylinder from a horizontal to a vertical position. The best way to lift a cylinder is to first make sure that the valve protection cap is secured tightly. Then raise the cylinder by grasping the cap firmly and lifting (Fig. 1-8).

Never allow oxygen cylinders to come in contact with live electrical wires or other electrical equipment. Keep cylinders away from welding and cutting work. Make certain that the hoses containing oxygen and acetylene do not lie under the work.

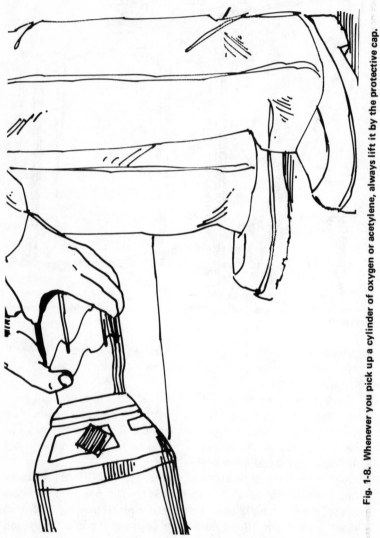

Fig. 1-8. Whenever you pick up a cylinder of oxygen or acetylene, always lift it by the protective cap.

Fig. 1-9. Always turn off the flow of both oxygen and acetylene at the tank when you stop work for more than 10 minutes.

Always close cylinder valves when you have completed working. Never leave the cylinder valve open when you are not in the immediate vicinity.   If you stop working for lunch, for example, turn off the valves and bleed the lines. I will explain this procedure later (Fig. 1-9).

Never use oxygen around oil or grease as these burn violently in the presence of pure oxygen under pressure. Do not use any oil or grease on the regulator fitting.

Never use oxygen for anything other than welding and cutting. Oxygen is flammable. There is a very real danger if it is used to "dust" off work or clothing. Similarly, oxygen should never be used for ventilation, pressure tests of any kind or for any other purpose.

ACETYLENE

Acetylene is produced by combining calcium carbide and water. When the calcium carbide, commonly called "carbide" in the trade, is dropped into water, a reaction occurs that causes gas bubbles to rise. The gas is acetylene. It has a peculiar odor and, if lighted, burns with a black, smoky flame. After the action of the carbide has ceased, a whitish residue remains in the water. This end product is hydrated or slaked lime (calcium hydroxide) and has many uses in the fertilizer industry.

It is interesting, as a side note, to mention that miners' lamps are called *carbide* lamps and work simply by adding water to carbide pellets. A gas is formed in a special container and is forced out of a jet. This gas is ignited and throws a bright beam (Fig. 1-10).

Carbide-to-Water Generators

Commercial production of acetylene is basically the same as a miner's head lamp, only on a much larger scale. Companies that use great amounts of acetylene will usually have their own generator which supplies them with all of the acetylene they can use. These generators are called "carbide-to-water generators" and are almost totally automated. In these types of generators, small amounts of calcium carbide are fed into a large sealed container of water. The heat given off as the reaction occurs is absorbed by the water. The acetylene gas is captured when it rises to the top of the tank. This gas is then either used directly or stored for future use (Fig. 1-11).

Containers

For the home welder, the only way to obtain acetylene is in a special tank, similar but much smaller than a standard oxygen cylinder. The three common sizes of acetylene cylinders are 300 cubic feet, 100 cubic feet and 60 cubic feet. A full tank of acetylene will register approximately 225 pounds per square inch on a pressure gauge (Fig. 1-12).

As with oxygen, the ICC has set up guidelines for acetylene containers. It is against ICC regulations to store free acetylene at pressures over 15 pounds per square inch. All welding and cutting with the oxyacetylene process can be done at pressures less than 15 pounds per square inch.

Fig. 1-10. A miner's carbide lamp works on the same basic principle as an acetylene generator.

In order to store acetylene below the 15 pounds per square inch (psi) requirement, special tanks have been developed. It is interesting to note how acetylene is stored. After a tank has been made from heavy gauge steel, it is packed with a porous substance such as pith from cornstalks, fuller's earth, lime silica or other materials. Next, the tank is filled with *acetone*, a liquid chemical having the ability of absorbing about 25 times its own volume of acetylene. When the cylinder is filled to about 225 psi with acetylene, the acetone dissolves the acetylene and makes it safe because the acetylene is not free (Fig. 1-13).

Acetylene cylinders have a valve on top to which the regulator is attached. This valve is similar to the valve on oxygen cylinders except it has left-handed threads. You will turn the valve counterclockwise to open and clockwise to close. This turning action is exactly opposite of a standard control valve.

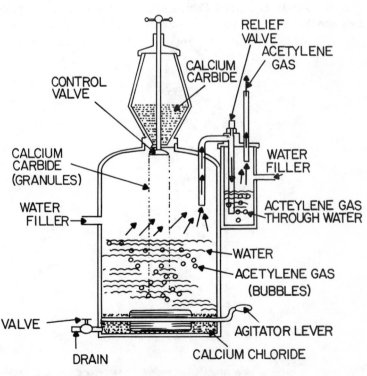

Fig. 1-11. Cutaway view of an acetylene generator.

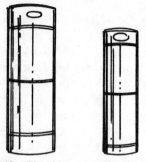

Fig. 1-12. The three most popular sizes of portable acetylene cylinders, left to right, are 300 cubic feet, 100 cubic feet and 60 cubic feet. Cylinders of other shapes are also available.

It should be noted at this time that all acetylene connections on the cylinder, hoses and torch handle are left-handed. Each brass fitting has a groove cut around its circumference for quick identification (Fig. 1-14).

All acetylene tanks have safety plugs on the top or bottom of the cylinder. These plugs are designed to melt, in the event of a fire, at around 212 degrees Fahrenheit. Needless to say, the cylinders should be stored and used away from any heat source. Once

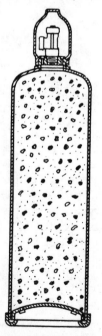

Fig. 1-13. A cutaway view of a typical acetylene cylinder showing the filler material.

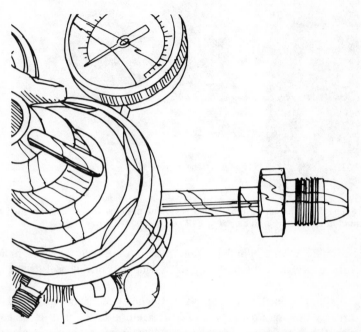

Fig. 1-14. The fitting is for an acetylene regulator. Notice that the nut has grooves around the middle. All left-hand threaded fittings have this groove.

these safety plugs melt, all of the acetylene and acetone in the cylinder is allowed to escape (Fig. 1-15).

Precautions

As with oxygen cylinders, there are a few general safety precautions which should always be followed. These include the following guidelines for safe operation and storage of acetylene.

Always call acetylene by its proper name — "acetylene." It should never be called "gas" as it is very different from the type of gas used in the kitchen, furnace or automobile.

Acetylene should be stored away from all heat sources and other fuels. Ideal storage conditions include a cool, dry area away from all combustible materials.

Handle acetylene cylinders with special care. The safety plugs melt at about the same temperature as boiling water and can "blow out" as a result of rough handling. As with oxygen, cylinders of acetylene should be secured while being used to eliminate the

Fig. 1-15. The safety plug on an acetylene cylinder is specially designed to melt at temperatures in excess of 212 degrees Fahrenheit.

possibility of the cylinder falling over. A specially designed hand truck is probably the best way of moving and storing acetylene and oxygen.

Do not open the valve on an acetylene cylinder more than 1 1/2 turns as this may cause some of the acetone in the tank to escape with the acetylene. Acetone will damage all rubber and plastic parts of the system, including the hose and regulator. Therefore, it should not be allowed to escape.

Always use acetylene in an upright position. Never use it when the cylinder is in any other position than vertical. Acetone may escape along with the acetylene.

Chapter 2
Oxyacetylene Welding Equipment

This chapter will examine the important parts of the oxyacetylene welding system. Proper use of the equipment, of course, results in better welding.

REGULATORS

As you know, oxygen and acetylene are commonly used from special cylinders. A full tank of oxygen will tip the scales at about 2,200 psi and acetylene will measure about 225 psi. Most welding and cutting can be done with a maximum of 50 psi for oxygen and 10 psi of acetylene. Obviously, some type of device must be attached to the tanks to reduce the high pressure of the tank to a lower working pressure. In addition, this device must also provide a steady flow of oxygen or acetylene over the life of the tank. The pressure in a full tank is obviously more than in a half-full tank, but the working pressure will be the same. Another requirement of such a device is that it must enable the user to regulate exactly how much working pressure goes into the torch. In short, this device must regulate consistently the flow of oxygen or acetylene from the tank to the torch.

There are basically two different types of regulators for the oxyacetylene welder: *single-stage* regulators and *two-stage* regulators. I will discuss both types.

Single-Stage Regulator

A single-stage regulator, as the name implies, reduces the pressure coming out of a cylinder to a desired working pressure in one step. These regulators are less expensive than two-stage regulators. They are practical for general welding, cutting and heating. A single-stage regulator will show more of a pressure rise or drop as the cylinder contents are used than does a two-stage regulator. Generally speaking, this deviation in pressure will not affect the quality of the

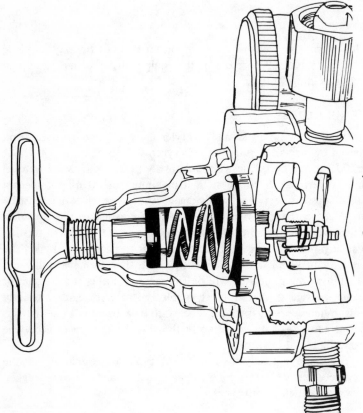

Fig. 2-1. Oxygen regulator.

flame on small, short-term work. But readjustments of the torch controls will be required when welding for a prolonged period (Fig. 2-1).

Two-Stage Regulator

A two-stage regulator will maintain practically constant flow pressures during a long period of operation. Two-stage regulators are more expensive than single-stage regulators, but offer much more precise flow regulation. Generally speaking, all two-stage regulators have two gauges; one shows the pressure in the tank and the other shows the working pressure in the line.

Actually the term two-stage means that the pressure in a tank is reduced in two steps before it is allowed into the line. The first stage serves as a high pressure reduction chamber, indicating the pressure in the tank. The second stage is, in effect, a low pressure reduction chamber and is adjusted by means of a special control knob on the face of the regulator.

There are many types of regulators being produced in this country. All of them, however, fall into either a single or two-stage classification. Many of these regulators have two gauges. One gauge indicates the pressure in the cylinder and the other indicates the working pressure being delivered to the torch (Fig. 2-2). Other

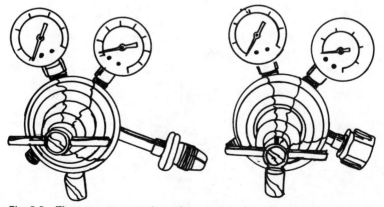

Fig. 2-2. These two stage regulators have gauges which show working pressure in the lines (left) and pressure in the tank. The regulator on the left is for acetylene and the other is for oxygen.

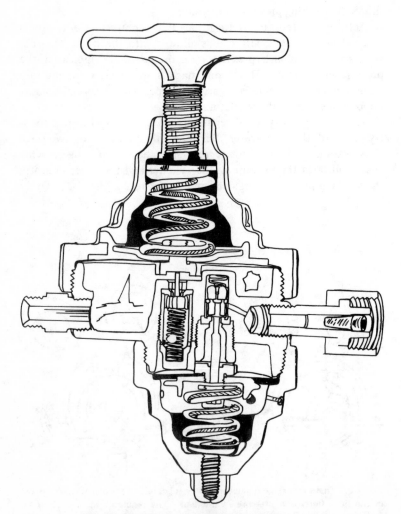

Fig. 2-3. Typical acetylene regulator without gauge.

regulators will have only one gauge, usually showing the working pressure being supplied to the torch. These single gauge regulators are most commonly single-stage regulators. There are also regulators that do not have any gauges. Generally these gaugeless regulators are used in commercial work, and are attached to a pipeline supplying oxygen and acetylene to a work station (Fig. 2-3).

Differences

Probably the greatest difference between a single and two-stage regulator is that a single-stage regulator will require adjustments to the torch as the pressure in the tank is lessened. One other objection to single-stage regulators is that they have a tendency to freeze in cold weather. A sudden expansion and resulting drop in initial pressure causes the gas to cool very rapidly. The end result is a frozen single-stage regulator.

There is a major difference between the regulator used for oxygen and one used for acetylene. Acetylene brass fittings are left-handed (and marked) and therefore cannot be used on oxygen tanks. Oxygen regulators are considerably stronger overall because of the pressure of standard oxygen tanks. A full tank will contain approximately 2,200 psi of pressure, while a full acetylene tank will contain only 225 psi.

Since regulators are probably the most important part of an oxyacetylene welding unit, they should be taken care of properly. An acetylene regulator is designed to handle and control the high pressure of acetylene and make using this gas safe. Oxygen regulators perform a similar function only at much higher pressures. Both types of regulators will provide years of dependable service if they are cared for according to generally accepted procedures. Although specific instructions will be included with all new welding regulators, some general comments about care can be given.

Care of Regulators

Never use any oil on an oxygen regulator as oil is easily ignited when it comes in contact with pure oxygen. This means that you should never oil any of the regulator parts (Fig. 2-4). Almost all regulators will have bold printing on their gauges stating "use no oil."

Before attaching a regulator to an oxygen or acetylene cylinder, crack the valve. Cracking means to open the valve slightly for a second. By doing this you will clear the cylinder outlet of any dirt that may have accumulated since the tank was last used. By re-

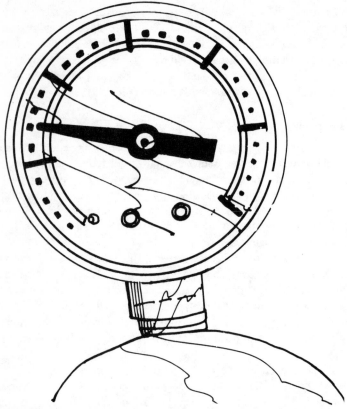

Fig. 2-4. Never use any lubricant on regulator parts. Most regulators say "use no oil" on the dial face as shown.

moving this dirt or dust, you will be insuring that these particles will not get inside the regulator where they can cause the regulator to malfunction (Fig. 2-5).

After the tank valve has been cracked, the regulator can be attached. Hand tighten, being careful not to cross thread the brass fitting. After the fitting is snug, tighten it with a wrench. With the regulator turned off, slowly open the valve on the cylinder. It is important that you do not open the cylinder valve too quickly. The pressure from the tank can damage the regulator if it is allowed into the regulator all at once. After opening the valve, check for leaks around the fitting.

Probably the best way to determine that the connections are snug is to use the soap test. Simply mix up a small amount of dish soap and water. Then put a few drops of this mixture on the exposed threads of the fitting. If a leak is present, the soap mixture will

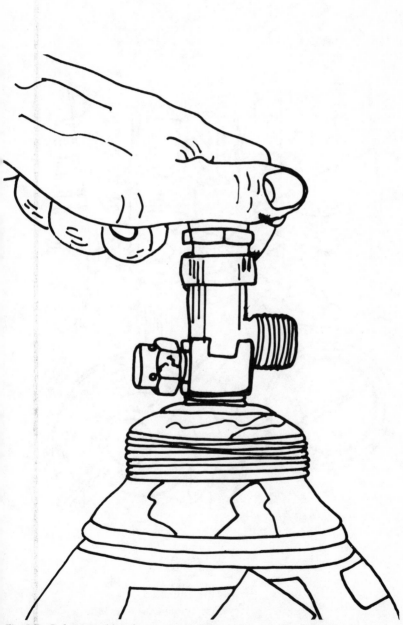

Fig. 2-5. Before attaching the oxygen regulator to the tank, the valve should be opened for a moment to clear the orifice. This is known as "cracking" the valve.

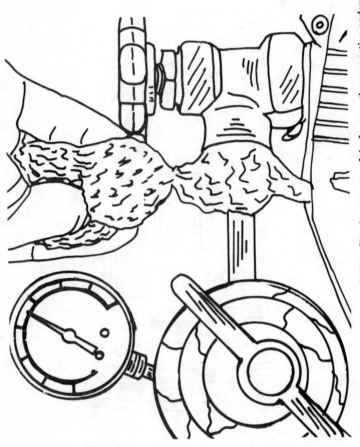

Fig. 2-6. To check for leaks, the soap bubble test is the most reliable. Simply put some of the foam on the connection and watch for large bubbles.

bubble. If the connection is tight, there will be no bubbles (Fig. 2-6).

Leaks around regulators are often the result of not tightening the fitting enough. In other cases, there may be a small particle of dirt on the regulator nipple, preventing a tight seal. Remove the regulator and wipe the fitting with a clean, dry cloth. Always turn off the tank valve before tightening a loose or leaking connection.

When regulators will not be used for a period of several weeks, you should relieve the pressure on the pressure valve seat. After the regulator has been disconnected from the cylinder, turn the pressure adjusting screw until it moves freely. In most cases this will mean screwing the handle out from the body of the regulator. The reason for doing this it to relieve the pressure on the valve seat and prolong the life of the regulator unit (Fig. 2-7).

The regulator valve seat should be kept clean at all times. Some regulator units come with plastic covers which are fitted over the regulator intake and outlet orifices. These do a lot to keep the regulator connecting fittings clean. In the absence of these covers, the threads of the fittings should be wiped clean with a dry cloth before attaching to the tank and fastening hose connections. While wiping these fittings, it is also a good idea to inspect them for signs of wear or damage to the threads. Damage to the threads will prevent a secure connection and the unit will leak. If this ever happens, the unit should be returned to the manufacturer or qualified welding equipment repair shop. Never use a regulator in need of repair.

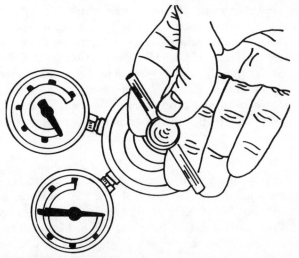

Fig. 2-7. To relieve the pressure on the regulator diaphragm, turn the control lever to the left until there is little resistance.

There are few problems with a well maintained regulator. It will provide years of dependable service if it is simply kept clean and not dropped or otherwise damaged. Nevertheless, there are a few things which may develop through general use.

Regulator Freezing

The first problem is freezing of the regulator. Since the function of a regulator is to control high pressure gas, the unit may become so cold that a layer of frost collects on the surface of the regulator. As the gas expands, it will absorb heat from the regulator and cool it. As the regulator becomes cooler, the moisture in the atmosphere condenses on the cold metal parts of the regulator and freezes.

Although frost on the outside of the regulator will not affect its operation, the gas flow may fluctuate, with a possible end result of the flame going out. Generally speaking, if the regulator freezes, it is a direct result of moisture in the gas and not a malfunction of the regulator.

Jump

Jump is fairly common regulator occurrence, but it is not really a problem at all. Jump usually happens when the flow of oxygen or acetylene is turned off at the torch handle. At this time, the gauge indicating the amount of pressure in the tank will rise momentarily and then stop. This short pressure rise is the jump. The amount of jump really indicates the true difference between pressure at the regulator outlet when oxygen or acetylene is flowing and when these gases are not flowing. When the torch valve is closed, pressure on the regulator's gauge rises until the diaphragm is moved far enough for the regulator valve to close tightly. This closed valve regulator is called *lockup*. Jump is not a regulator malfunction. It is a normal reaction to changes in flow or no flow pressures.

Creep

Creep is probably the only true problem with regulators. Creep can happen when the torch valve is shut off at the torch handle. The low pressure or working pressure gauge then moves upward. Limited creep is very similar to jump and may be hard to distinguish. In extreme cases, creep will cause the gauge to rise until either it registers the same pressure as the cylinder or a connection is blown, usually in the hose.

Creep always occurs when the regulator is defective and cannot reach lockup. Needless to say, you should never operate a regulator which has a tendency to creep.

HOSES

Hoses are another important part of the oxyacetylene welding system. There are always two hoses, one for oxygen and another for acetylene. Each end of the hose has a fitting which attaches to either the regulator or the torch handle. As with all other connections, the oxygen fittings are right-handed threads and the acetylene connections are left-handed with an identifying groove cut around the circumference of the brass fitting. All hoses are colored — green for oxygen and red for acetylene. Additionally it is common practice, at least with new equipment, for the hose maker to print "oxygen" on the green oxygen hose. It is impossible to attach the acetylene hose to the oxygen fitting on either the regulator or torch handle because the threads, as mentioned, are different (Fig. 2-8).

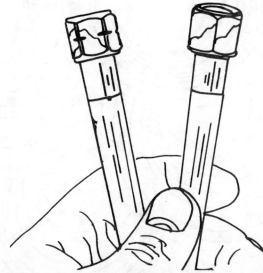

Fig. 2-8. To relieve the pressure on the regulator diaphragm, turn the control lever to the left until there is little resistance.

Generally speaking, all hoses being produced today are made from a high quality synthetic or natural rubber material. Oxygen and acetylene hoses are flame and oil-resistant and are also reinforced with a nylon mesh inside the hose. The end result is a very tough hose that is tested at the factory to withstand pressures of up to 400 pounds per square inch (Fig. 2-9).

Most of the welding hoses in use today are double or *Siamese*, as they are called in the trade. This simply means that the two hoses are joined along their entire length with a special adhesive. Handling the hose is much easier than if two separate hoses were being used. Hoses for welding are available in several different lengths and inside diameters, the most common being 20 feet long and 3/16 inside diameter.

A quality welding hose will last for many years with a minimum of care. Before using a new hose, the talcum powder, which all new hoses are lined with, should be blown out with compressed air. Failure to do so may cause some of this powder to enter the regulator or torch handle where it can cause a malfunction. Probably the easiest way to do this is to bring the new hose to your local gas station and use the air hose to blow out the lines.

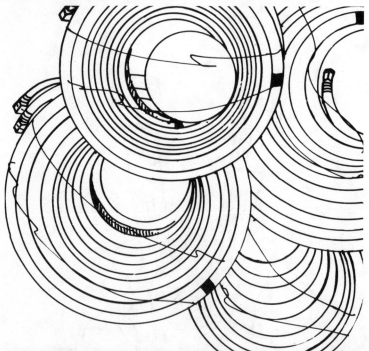

Fig. 2-9. Hoses for welding are specially designed to withstand high pressures.

Check your welding hose periodically for signs of wear or deterioration. It is a good idea to wipe the hoses regularly with a clean, dry cloth. In cases of extremely dirty hoses, you can wipe them with a damp cloth. As you are cleaning the hoses, check for signs of wear (Fig. 2-10).

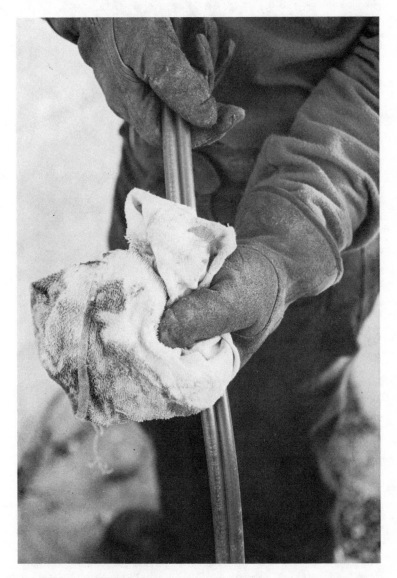

Fig. 2-10. It is a good idea to wipe oxygen and acetylene hoses periodically to keep them clean.

Fig. 2-11. Check your hoses for leaks by submerging them in a pail of water with pressure in the lines.

Older hoses can be checked for leaks by immersing them in a pail of water while they are under working pressure. The same soap and water test mentioned earlier can be used to check connections around fittings and other parts of equipment under pressure. If a leak is ever discovered it should not be repaired with tape, but should instead be taken to a qualified repair shop (Fig. 2-11).

While working, it is a good idea to never step on the hose. You should also make certain that the hoses never lie where they can be damaged by hot sparks or slag. It is a good practice when storing welding hoses for a long period of time to put a piece of tape over the fittings to prevent the entrance of dirt or insects. Masking tape works well for this and will insure that your hoses remain internally clean and clear of obstructions which could damage equipment (Fig. 2-12).

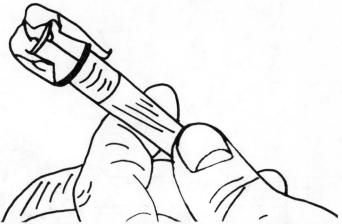

Fig. 2-12. When storing hoses for a prolonged period, it is a good idea to cover the ends with a piece of tape. This will keep dust, dirt and insects out of the hoses.

TORCHES

The business end of any oxyacetylene welding outfit is the *torch* or, to use the correct terminology, the *blowpipe*. The oxygen and acetylene pass through their respective regulators, through the hoses and into separate fittings in the handle of the torch. A control knob for each regulates the flow through the torch handle and, in fact, controls the flame (Fig. 2-13).

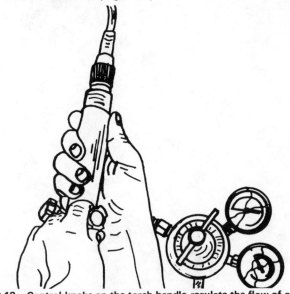

Fig. 2-13. Control knobs on the torch handle regulate the flow of oxygen and acetylene.

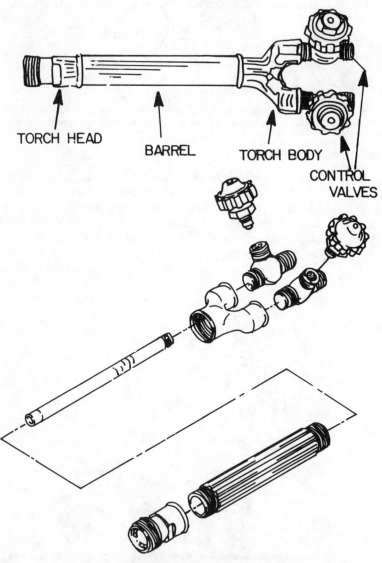

Fig. 2-14. Diagram of a welding blowpipe.

TORCH HEAD

BARREL

TORCH BODY

CONTROL VALVES

Fig. 2-15. One of the many complete welding and cutting projects.

At the present time there are two types of torches currently available: *injector* and *medium pressure*. The major difference between the two is that the injector type can use acetylene at pressures less than 1 pound per square inch while the medium pressure torches require that the acetylene be supplied in pressures from 1 to 15 pounds per square inch. We will discuss both types of blowpipes (Figs. 2-14 and 2-15).

Injector Blowpipes

Injector type blowpipes, often called low pressure blowpipes, are all similar in principle but will differ slightly from one manufacturer to another. Generally speaking, they all work as follows. The rear

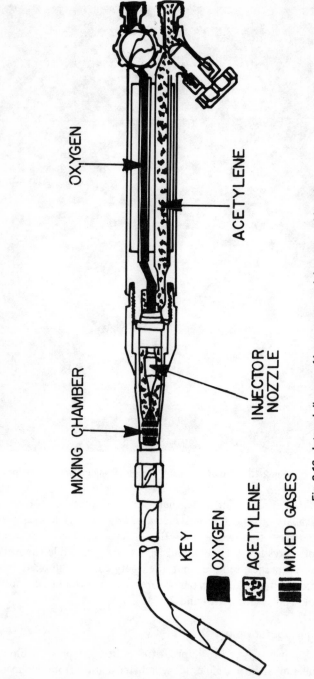

OXYGEN

ACETYLENE

MIXING CHAMBER

INJECTOR NOZZLE

KEY
- OXYGEN
- ACETYLENE
- MIXED GASES

Fig. 2-16. Internal diagram of low pressure or injector type blowpipe.

of the torch handle has two hose connections, one for oxygen and another for acetylene. The front of the handle forms a chamber where the oxygen and acetylene meet for the first time and are mixed before passing through the tip to be burned.

As oxygen passes through its tube in the handle and into the mixing chamber, it creates a suction action which draws acetylene along and into the mixing chamber as well. Here the oxygen and acetylene are allowed to expand slightly and to be thoroughly mixed. As the name low pressure implies, these blowpipes have the ability of working under very low acetylene pressures and are handy for light duty work such as wire sculpture. They are also commonly used in industry where acetylene is supplied from a generator. Low pressure blowpipes can also be used with cylinders of oxygen and acetylene as well (Fig. 2-16).

Medium Pressure Blowpipes

Medium pressure blowpipes, as mentioned earlier, require acetylene with a minimum pressure of 1 pound per square inch and, up to 15 pounds per square inch. The major difference between a medium pressure blowpipe, sometimes called a balanced pressure blowpipe, and a low pressure blowpipe lies in how the oxygen and acetylene are mixed. Although specific details of this function will differ between manufacturers, they all operate in the same basic manner. Inside the blowpipe handle there will be a large hole in the center surrounded by several smaller holes. All of these holes are located just behind the mixing chamber and base of the tip. Oxygen passes through the center hole and acetylene exits through the surrounding smaller holes. Some medium pressure blowpipes are exactly opposite, but the basic principle remains the same. The gases are then allowed to expand momentarily so thorough mixing can take place before passing out the tip to be burned. In all cases medium pressure blowpipes are designed to operate at equal pressures of both oxygen and acetylene. The manufacturer of the blowpipe will always supply instructions and recommendations for both oxygen and acetylene pressures so that a specific flame can be developed at the tip (Fig. 2-17).

Medium pressure blowpipes are probably the most common type of welding torch in use today. They provide as versatile welding tool as possible within the limitations of oxygen and acetylene (Fig. 2-18).

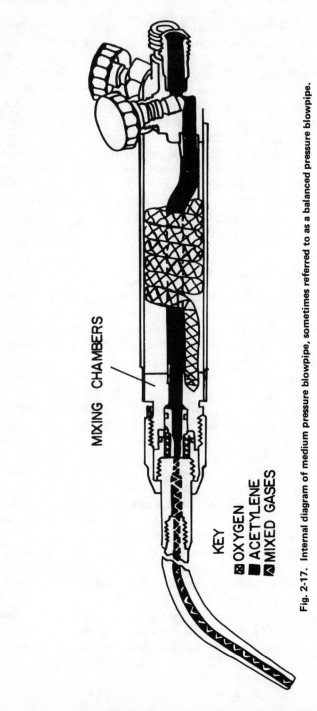

MIXING CHAMBERS

KEY
⊠ OXYGEN
■ ACETYLENE
⧄ MIXED GASES

Fig. 2-17. Internal diagram of medium pressure blowpipe, sometimes referred to as a balanced pressure blowpipe.

Fig. 2-18. Medium pressure blowpipes are probably the most common type for general purpose welding.

TIPS

No welding blowpipe is complete without a welding tip, and there are many different sizes. Each torch maker offers a line of various tips designed to be used with their particular unit. Basically all welding tips are made the same. They are most commonly made of a heavy gauge copper tubing which has a high resistance to reflected heat. Some tips are straight while others are curved. In principle, all welding tips are simply a piece of tubing with a threaded

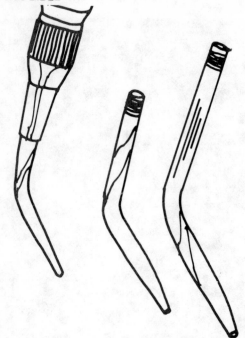

Fig. 2-19. Different size welding tips can be used with the same torch handle.

end which attaches the tip to the torch body. The other end has a hole through which the oxyacetylene mixture flows and is burned (Fig. 2-19).

All welding torch makers offer tip size suggestions for the various thicknesses of metals being welded with their particular blowpipe. Basically, the thickness of the metal will determine the tip size. The tip size being used will have a bearing on the pressure settings for both oxygen and acetylene (Table 2-1).

Table 2-1. Pressure Chart for Welding Tips.

Metal Thickness	Tip Size	Size of Welding Rod (inches)	Oxygen PSI	Acetylene PSI
1/32	No. 1	1/16	5	5
3/64	No. 2	1/16	5	5
1/16	No. 3	1/16	5	5
3/32	No. 4	3/32	5	5
1/8	No. 5	3/32	5	5
3/16	No. 6	3/32	6	6
1/4	No. 7	1/8	7	7
5/16	No. 8	5/32	8	8

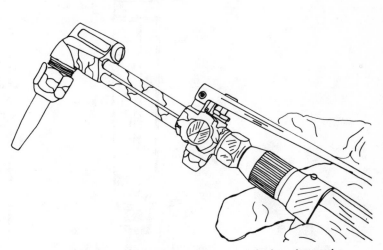

Fig. 2-20. The oxygen cutting lever on a standard cutting torch.

Most welding blowpipes have a cutting attachment which fits on the torch handle in place of a welding tip. Cutting blowpipes (often called the head of an oxyacetylene blowpipe) can be either low pressure or medium pressure depending, of course, on the basic type of unit. All cutting blowpipes are designed to allow mixed oxygen and acetylene to be released through special preheat orifices in the head of the unit. All cutting tips are the same, though the holes may be of different diameters. Basically, there are a series of small holes surrounding a larger hole in the center of the tip. The oxygen and acetylene mixture burns at the smaller holes. The center hole allows a stream of pure oxygen to be added to the flame by the welder simply by pressing the long lever on the cutting attachment (Fig. 2-20). It is this flow of pure oxygen that does the actual cutting after the metal has been preheated by the flame (Figs. 2-21 and 2-22).

Cutting Attachments

Cutting heads can be either the injector or medium pressure type. Obviously only the proper type of cutting attachment should be used with the torch handle you have. A low pressure cutting head

Fig. 2-21. The cutting attachment fits into the torch handle in place of welding tips.

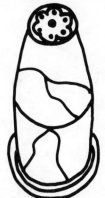

Fig. 2-22. A standard cutting attachment tip has six preheat holes surrounding a larger oxygen cutting hole in the center.

is attached to a low pressure blowpipe, for example. The operation of both types is the same as for the respective welding handles covered previously.

Just as there are different size welding tips, there are also different size cutting tips. Each torch maker supplies recommended cutting tip sizes for various thicknesses of metal. Oxygen/acetylene pressures should be adjusted accordingly (Table 2-2).

Slag and Flashback

We have discussed both welding and cutting tips in this chapter. This will be a good time to mention a few things that should be done to all tips. Because the cutting or welding tip is necessarily very close to the work, it may become clogged with *slag*, tiny bits of molten metal. When this happens the flame will sputter and, in extreme cases, possibly go out. The only solution when the flame begins to spit and spurt is to clean the tip with either a special tip cleaning tool or a suitable size drill bit. It is important, when cleaning any welding or cutting tip, to move the tool in a straight in and out motion so the hole is not enlarged or made bell-shaped (Fig. 2-23).

Table 2-2. Pressure Chart for Cutting Tips.

Metal Thickness (inches)	Tip Size	Oxygen PSI	Acetylene PSI
3/8-5/8	No. 0	30-40	5-15
5/8-1	No. 1	35-50	5-15
1-2	No. 2	40-55	5-15
2-3	No. 3	45-60	5-15
3-6	No. 4	50-100	5-15

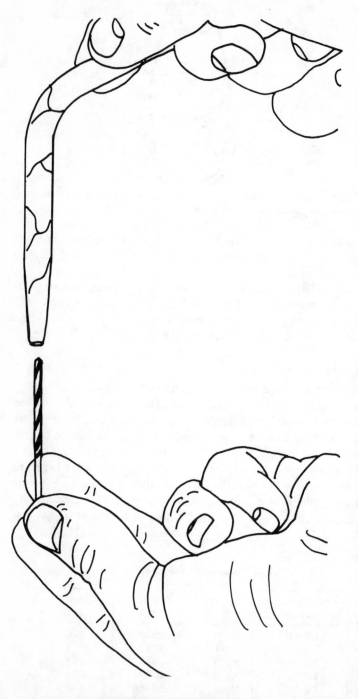

Fig. 2-23. Welding tips will become clogged with normal use. While cleaning tools are available, the proper size twist drill bit works just as well if it is used properly.

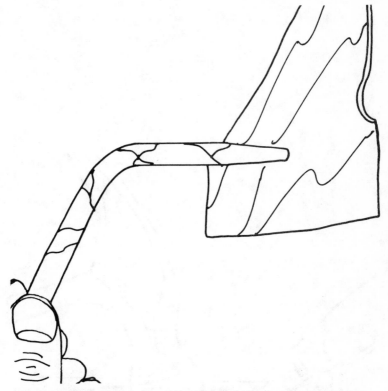

Fig. 2-24. Clean the end of a welding tip by rubbing it across emery paper.

Often, especially when the tip of the blowpipe is placed too close to the work, the face of the tip, in addition to becoming clogged, will form a buildup of metal. This buildup should be removed with an abrasive — a file or emery cloth. It is important to work carefully to remove carbon and slag deposits whenever they happen (Fig. 2-24).

If a welding or cutting tip should ever become damaged, as a result of bending or cross threading, it should be replaced. Usually a small bend can be repaired by simply placing the tip on a block of wood and striking gently with a plastic or leather mallet. Care must be exercised when doing this, however, so the copper tubing is not compressed (Fig. 2-25).

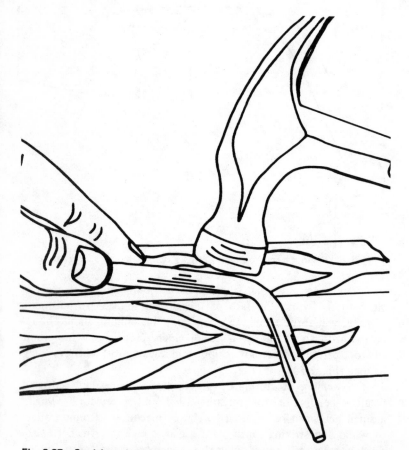

Fig. 2-25. Straighten bent welding tips with light blows from a hammer. Use a wooden block underneath.

Fig. 2-26. Flashback arrestors are attached to the torch handle and eliminate the possibility of a flashback. Most new torches come with these special safety fittings.

If a torch tip should become clogged, as a result of slag buildup, the burning oxygen and acetylene may flow back into the hose and regulator causing what is known as a *flashback*. Special fittings have been developed to reduce the danger of a flashback. If your torch does not have them, you should consider the investment.

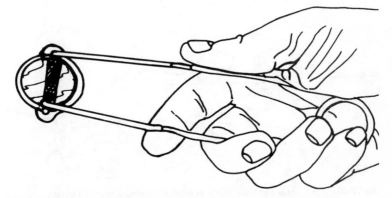

Fig. 2-27 The standard striker supplied with most welding supplies.

These fittings are usually attached between the end of the hoses and the torch handle. They will close off the flow of oxygen and/or acetylene whenever there is a flashback and the flow of gas is reversed. Many of the new welding outfits on the market today have these special fittings included with the unit (Fig. 2-26).

CLOTHING AND GEAR

In addition to oxygen and acetylene tanks, regulators, hoses, torch, and an assortment of tips, the do-it-yourself welder will need some safety clothing and gear including goggles (with lenses designed for gas welding), leather gloves, long-sleeved shirt, heavy pants without cuffs and, possibly, a hat. Protective clothing is covered in detail in the chapter on safety, it will not be dealt with here. One other item the welder will need is some type of ignition system to light the torch. Matches are definitely out of the question.

Probably the most common type of torch lighter is the one shown in Fig. 2-27. In theory, this lighter produces a hot spark every time the flint is rubbed across the file in the bottom of the cup. In practice, however, this type of torch lighter is not worth the price of the steel from which it is made. Unfortunately most welding outfits — regulators, hoses, torch, cutting head and tips — usually come with one of these lighters. In my opinion, the first thing you should do when you purchase your welding outfit is to throw out the striker that comes with it (if it looks like the one in Fig. 2-27 and buy a good one instead. Welding supply houses, the same places where you buy your acetylene and oxygen, will often carry several different types of strikers.

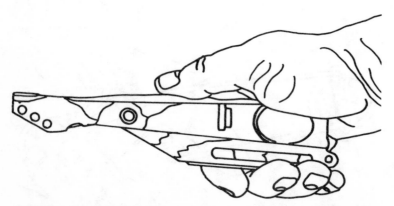

Fig. 2-28. This new type of torch lighter produces about 10 bright sparks every time the handle is squeezed.

One relatively new addition to the lighter line is an electronic striker. This type does not rely on the old principle of a piece of flint and steel but instead uses a tiny hand-operated generator to throw a spark between two terminals. The heart of this little unit is a tiny magneto, similar to the magneto in old time Model "T" Fords and older tractors. Everytime the handle is squeezed, a series of sparks jump between the terminals. The result is that the torch can be quickly ignited (Fig. 2-28).

Chapter 3

Setting Up an Oxyacetylene Outfit

Since the oxyacetylene welding process is the most popular and versatile means of joining or cutting metal for the do-it-yourselfer, it will be helpful to explain how to set up the equipment. In this chapter we will cover some of the basics of torch adjustment and standard safety practices. More specific information such as how to braze, weld or cut metal can be found in later chapters.

Fig. 3-1. A cylinder hand truck is probably the best way to securely store or move oxygen and acetylene tanks.

Fig. 3-2. Crack the oxygen valve.

ATTACHING REGULATORS AND HOSES

To set up for oxyacetylene welding, we begin by making certain that the tanks of oxygen and acetylene are securely fastened to an immovable object, such as a pole or a cart specifically designed to hold the tanks. Welding supply houses all seem to carry a line of hand trucks for this purpose. They are a good investment because they safely hold the tanks and give the welder mobility as well (Fig. 3-1).

Next, the regulators must be attached to the tanks. First, the oxygen tank is cracked. The valve is opened slightly so some of the oxygen escapes and clears the valve orifice. The oxygen regulator fitting is wiped with a clean dry cloth and then fitted into the valve on top of the oxygen tank. The regulator fitting is tightened by hand. Then a special wrench is used to make sure the connection is snug. After the oxygen regulator has been secured, the acetylene regulator is attached to the acetylene cylinder valve in the same manner (Figs. 3-2, 3-3 and 3-4).

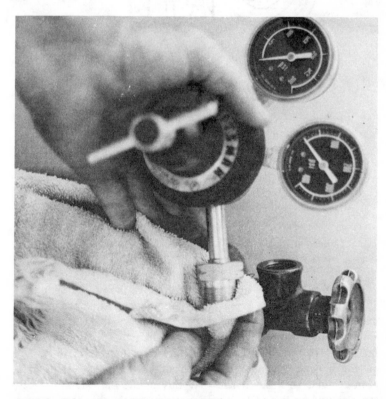

Fig. 3-3. The regulator fittings should be wiped with a clean dry cloth before attaching to the tank valve.

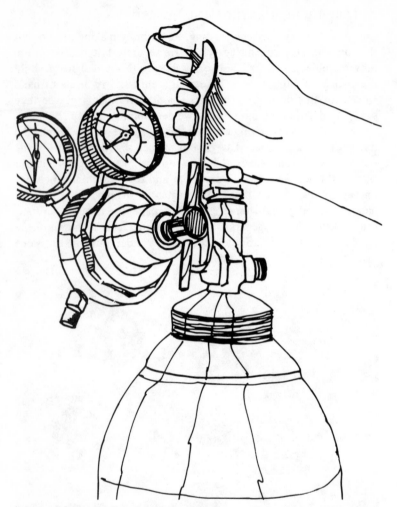

Fig. 3-4. After the regulator fitting has been snugged up by hand, tighten it securely with a wrench.

After the regulators have been attached to the proper cylinders, attach the hoses: the green one to the oxygen tank and the red one with the left-handed threads to the acetylene tank. If the hoses are new, they should have been blown clear with compressed air first. Attach the other ends of the hose to the respective fittings on the blowpipe handle. Make certain that all connections have been secured with a wrench. At this point, the valves on both tanks should be turned to the off position, as should the regulators and the controls on the torch handle (Fig. 3-5).

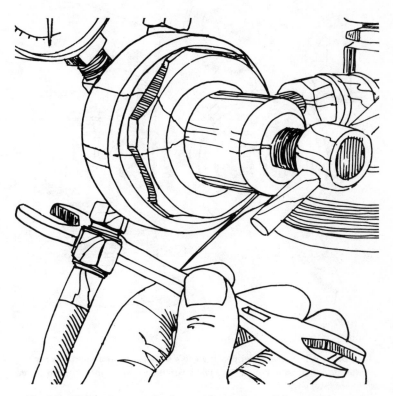

Fig. 3-5. Tighten hose connections to the regulator with a special wrench.

OPENING VALVES

The next step is to very slowly open the oxygen cylinder control valve, letting the oxygen into the regulator. You must do this slowly so there will not be a surge of pressure on the regulator diaphragm. Once the gauge begins to register pressure, open the oxygen valve one to 1 1/2 turns. Then open the control valve on the acetylene tank. This valve turns to the left and should also be opened slowly, not more than one full turn (Fig. 3-6).

After all connections are tight and the oxygen and acetylene have been turned on, you should give a quick check to make certain that there are no leaks. Since you will not be able to see leaks, you must check by listening or smelling all of the connections. If you are ever uncertain as to whether or not a connection is leaking, use the soap test described earlier in this chapter.

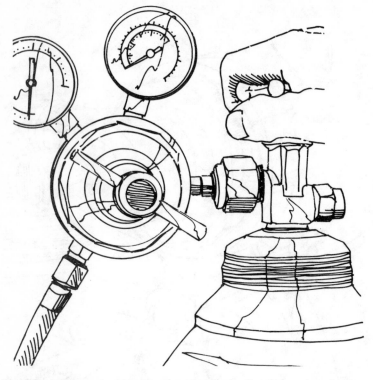

Fig. 3-6. It is important to open the valve slowly after the regulator has been securely fastened to the tank. This will insure that there is not a surge of oxygen to the internal parts.

SETTING WORKING PRESSURES

After you are sure that all connections are secure, you can set the working pressures in the hoses. This is done by turning the control levers on the face of the regulators. You will have to consult the chart which came with your welding outfit to determine the working pressure for the tip being used (Fig. 3-7).

While setting the working pressure, it may be necessary to open the corresponding control knob on the torch handle. This will clear the lines and enable you to check that oxygen and acetylene are, in fact, flowing through the handle. After you have set the working pressures — work on one line at a time — turn off the control knob on the handle and watch the regulator gauges. They should only move slightly if at all. You must set the working pressure on both the oxygen and acetylene according to the tip chart for your outfit.

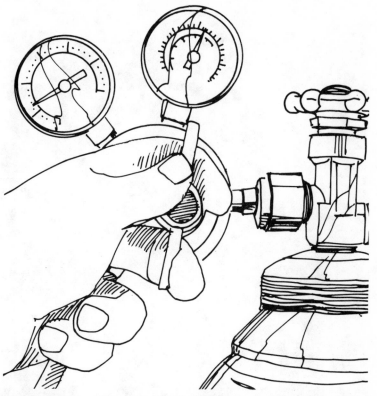

Fig. 3-7. Set the working pressure in the hose by turning the regulator control lever to the right. The oxygen working pressure is being set at about 30 psi while the pressure in the tank is about 1700 psi.

When setting the pressure in the hoses, with the gas or oxygen flowing into the atmosphere, you must make certain that there is no open flame or other heat source in the area. This could cause the fuel to ignite.

LIGHTING THE TORCH

After the proper working pressures have been set, you can light the torch. Before you do, however, you should put on a pair of gloves and goggles. You can raise the goggles to your forehead, so they can be easily pulled down over your eyes when you need them (Fig. 3-8). You should also be standing in front of your workbench, with the hoses coming from behind you so there is no possibility of them being hit and damaged by a hot spark or slag. Assuming that you are right-handed, hold the blowpipe in this hand and

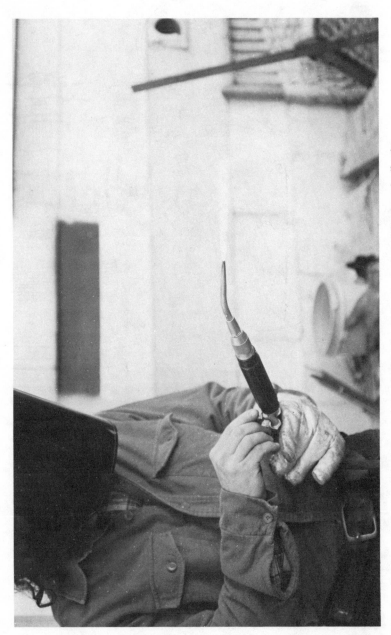

Fig. 3-8. Proper eye protection should be worn when lighting and adjusting the torch.

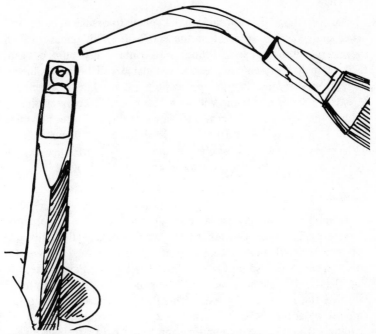

Fig. 3-9. Hold the striker close to the torch tip, with the acetylene flowing. Light the gas by striking a spark into the gas.

the striker in your left. Turn the acetylene control knob open a crack, just enough to let some of the gas flow through the tip. Hold the striker close to the tip and squeeze it a few times to generate enough sparks to light the acetylene (Fig. 3-9).

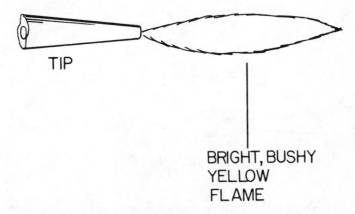

TIP

BRIGHT, BUSHY
YELLOW
FLAME

Fig. 3-10. A pure acetylene flame containing no oxygen.

Acetylene Flame

The first flame that appears will be yellow to orange in color, and black smoke will be present at the end of it. This flame is called the *acetylene flame* and it does not contain any pure oxygen. If too much acetylene is present, the flame will be a few inches away from the hole in the tip, possibly roaring like a jet engine. In extreme cases this flame will blow itself out. To correct, simply reduce the amount of acetylene flowing through the blowpipe handle. When the acetylene flame is about 8 to 10 inches long, with only a small amount of black smoke, begin to introduce oxygen by slowly opening the oxygen control knob (Fig. 3-10).

Carburizing Flame

As soon as pure oxygen is added to the acetylene flame, some changes begin to happen. First the black smoke will totally disappear and the flame will change from a yellow/orange to a whitish color. There will also be three distinct parts to the flame as seen in Fig. 3-11. The flame produced when oxygen is first introduced is called the *carburizing flame.* This flame, because of the greater amount of acetylene in relation to oxygen is not hot enough for welding. It can be used to add carbon to a weld or to blacken the surface of the metal.

Neutral Flame

As more oxygen is added to the flame, simply by turning the oxygen control knob, will change once again. The excess acetylene feather will appear to shrink towards the tip and will finally disappear

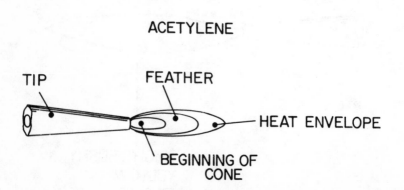

Fig. 3-11. As soon as oxygen is introduced to the acetylene flame, a cone begins to form and the flame takes on a definite shape. It is called a carburizing flame.

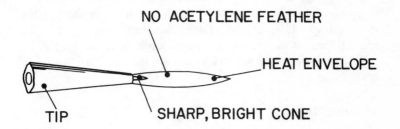

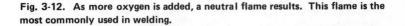

Fig. 3-12. As more oxygen is added, a neutral flame results. This flame is the most commonly used in welding.

into the small white cone at the base of the tip. This flame is called the *neutral flame* and is used to weld steel and braze most metals. There are only two parts to this flame, a sharp bright cone at the end of the welding tip and a heat envelope (Fig. 3-12). The cone should be well defined and white in color.

The name neutral flame is derived from the fact that the chemical effect of this flame on molten metal during the welding process is neutral. This is true, of course, providing that the flame is held in the proper manner with the inner cone not quite touching the

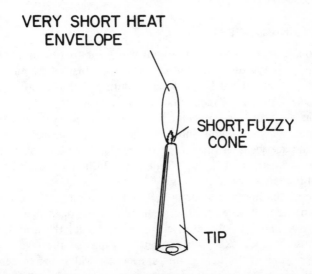

Fig. 3-13. If more oxygen is added to the neutral flame, the cone will decrease in size. The result is an oxidizing flame.

molten metal. It will be to your advantage to learn how to adjust your blowpipe to arrive at a neutral flame, as it will be the most frequently used flame in your work. You can adjust the length of the cone by adding more acetylene and then more oxygen. The neutral flame is important not only because it is the most widely used flame, but it also serves as a basis for adjusting to or from other flames.

Oxidizing Flame

If the oxygen control knob is opened more, the nature of the neutral flame will change once again. The new flame is called an *oxidizing flame.* This flame is often used for brazing. To adjust a neutral flame to an oxidizing flame, you must increase the ratio of oxygen to acetylene. You can either reduce the amount of acetylene coming out of the blowpipe or increase the amount of oxygen (Fig. 3-13).

An oxidizing flame should never be used for welding. The excess oxygen will cause breakdowns and defects in both the metal being welded and the weld itself.

When you first begin welding with an oxyacetylene outfit, you should practice adjusting the blowpipe to the different types of flames: acetylene, carburizing, neutral and oxidizing. Once you have become proficient at adjusting to these flames, especially the last two, you can then work on technique.

CONCLUDING TASKS

After you have finished welding, you will want to turn off your blowpipe. Always turn off the acetylene flow first. Oxygen will continue to flow through the torch, insuring that the orifice is blown clean. After a few moments, turn off the oxygen.

If you are stopping work for just a short period of time, say 10 minutes or more, you should close the tank valves, leaving oxygen and acetylene in their respective hoses. If you are stopping work for a longer period you should, in addition to closing the valves on the tanks, bleed the lines.

To bleed the hoses in your unit, begin by turning off the flow of oxygen and acetylene into the regulators. Then open the acetylene control knob on the blowpipe handle and let the acetylene in the hose escape. Keep an eye on the gauge of the regulator. As the pressure in the hose drops, the needle will swing to zero. Once this happens close the control knob on the handle and repeat the operation, bleeding the oxygen line.

Fig. 3-14. After the hoses have been disconnected, it is a wise practice to turn the regulators off by turning to the left. This will save wear and tear on fragile internal parts.

Another task you must accomplish when shutting down your welding outfit, after the hoses have been bled, is to release the regulator pressure handles so there is no tension on the diaphragms. Do this simply by turning the handle on the face of the regulator until there is very little or no resistance. This small act will go a lot to prolonging the life of your regulators (Fig. 3-14).

Chapter 4
Other Types of Welding and Equipment

Up to this point, we have covered only the equipment necessary for oxyacetylene welding and cutting. The bulk of this book will, in fact, concern itself with this method of joining metals. I believe it is the easiest type of welding for the beginner to learn and it is versatile enough to enable the do-it-yourselfer to perform most types of metal joining tasks. Remember that all welding falls into three rather broad categories: gas welding, electric welding and gas/electric welding. Since we have already covered the equipment necessary for the first type, it is now time to turn our attention to the others (Fig. 4-1).

Fig. 4-1. Arc welding in progress.

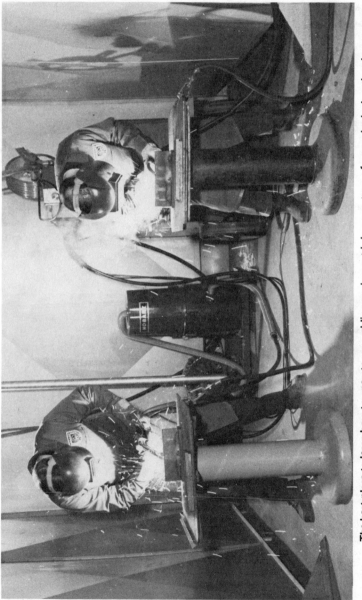

The best place to learn how to operate arc welding equipment is in some type of supervised educational program.

ARC WELDING

If you feel that you would like to try your hand at *arc* welding, I suggest that you contact a local welding supply house for more information and possibly a demonstration. There are also courses offered at vocational schools and adult education classes. In truth, arc welding is not difficult but it does require a certain discipline and strict adherence to specific safety practices. Learning the basics of arc welding is best accomplished under the tutelage of a knowledgeable expert (Fig. 4-2).

Shielded metal arc welding, commonly referred to as simply arc welding, is probably the most widely used form of welding in use today. It is extremely attractive for commercial work because it is fast and produces strong welds. The major application of arc welding is for joining mild carbon and low alloy steels.

Electrodes

The underlying principle of arc welding is quite simple. An electric welding machine produces a specific amount of current which passes through a cable, called an *electrode lead*, to a special handle which holds an electrode. An electrode is similar to a welding rod used for gas welding except that it must also conduct electricity. Electrodes are also covered with *flux*. An arc is struck by moving the electrode close to the metal. The arc is extremely hot, enough to melt both the end of the electrode and the edges of the metal. The current passes through the electrode, melting the end and the metal where the arc from the electrode touches the metal. A work-lead cable connects at the welding machine and is also clamped to the metal being welded, making it possible for the electric current to complete its circular path (Fig. 4-3).

The arc creates a very intense heat which melts the edges of the metal and the end of the electrode at the same time. The melted metal from the electrode moves across the arc and is deposited on the metal forming a strong weld, in most cases equal to or stronger than the base metal itself.

The flux coating on the electrode when an arc is struck produces an inert gas that shields the arc and the weld from the surrounding atmosphere. At the same time deoxidizers are produced which purify the electrode metal. This flux also forms a slag, which protects the molten metal from oxidation. After the weld has cooled a bit, the slag is commonly removed with a special tool (Fig. 4-4).

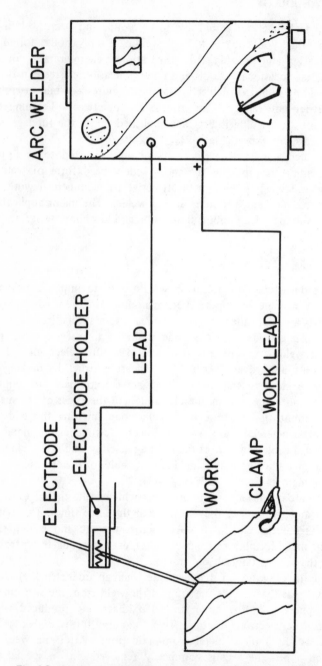

Fig. 4-3. An oversimplified diagram of a typical arc welding unit.

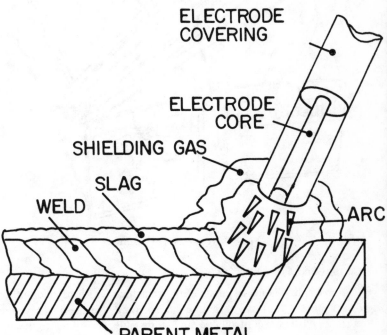

Fig. 4-4. As the electrode covering melts, a special shielding gas is produced. This gas protects the molten metal from the atmosphere.

As you can well imagine, electrodes are a very important part of an arc welding system. They are commonly identifiable by a special numbering system developed by the American Welding Society according to strength, welding position, usability and analysis of the deposited metal (Fig. 4-5).

Probably hundreds of different flux-coated electrodes have been developed to perform under various applications. Electrodes are available in sizes ranging from 1/16 to 5/16 inch diameter and from 9 to 18 inches long. Fourteen-inch ones are probably the most widely used in the industry.

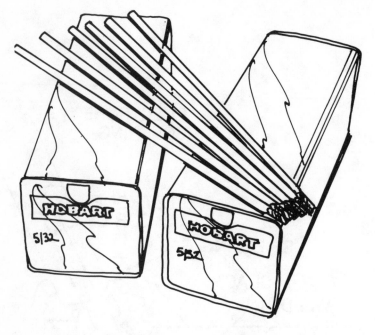

Fig. 4-5. A selection of electrodes.

Current-Supplying Machine

Arc welding is made possible by a machine which supplies either direct current (dc) or alternating current (ac). Needless to say, the current must be controllable and adjustable according to needs. At one time all arc welding was done with dc because ac was unsuitable. Now, however, many welding equipment companies have come out with a line of ac welding units that will weld a wide range of thickness and types of metals (Fig. 4-6). One advantage of ac arc welding units is that they can be produced at a lower cost, which makes these units affordable by more potential arc welders. The ac transformer type is probably the smallest, lightest and least expensive of all. It is also relatively quiet during operation.

An arc welding machine, either ac or dc, provides the electric power in the proper current voltage to maintain the welding arc. The higher the amperage, the greater the electric flow that will be transferred through the electrode to the metal being welded. Also, the higher the current, the faster the electrode will be consumed. Generally speaking, the higher settings on the welding machine are reserved for the thicker base metals being welded.

Fig. 4-6. A typical arc welding machine.

Protection Gear

Arc welding requires the welder to wear a special helmet to protect the eyes from the extremely bright arc and to protect the face and head from flying particles. It is also standard practice to wear heavy gauntlet gloves and heavy clothing. The clothing and protection gear is often bulky and confining, but it is necessary to provide a line of defense against burns and eye damage (Fig. 4-7).

While it is not my intention to cover arc welding in detail, I would like to mention some of the basics of this form of metal joining. The first step to be learned is to strike a smooth arc without

Fig. 4-7. Standard arc welding face mask.

allowing the electrode to touch the metal. When this happens, the electrode will stick to the metal. This is known as *freezing*. If this ever happens there are two courses of action, one of which should be initiated immediately.

Striking an Arc

The first thing you can try when freezing of the electrode happens is to give a firm twist to the electrode holder. If a fast twist with an upward pull does not free the electrode, release the electrode from the holder and break the circuit. After the metal and electrode have

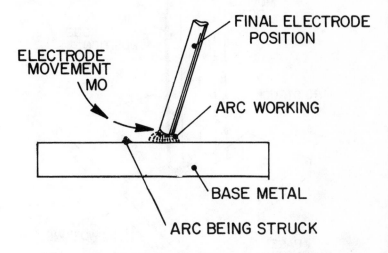

Fig. 4-8. Use the scratch method to strike an arc with an ac arc welder.

cooled, the electrode can easily be removed from the metal with a pair of pliers, using a twisting motion. Failing this, try striking the base of the electrode, where it is joined to the metal with a chipping hammer.

Because of the possibility of freezing and the potential damage to equipment, it is a sound idea to practice striking arc a few times with the power turned off. Once you have the general idea, then you can proceed with the power on.

To strike a proper arc, begin by setting the recommended current in the machine. Then fasten an electrode in the holder. Hold the electrode almost straight above the metal to be welded and tilted at about a 10 degree angle. If you are using an ac machine, you strike an arc using the scratching method (Fig. 4-8). This is similar to striking a kitchen match. If you are using a dc machine, the accepted method of striking an arc is to tap the electrode to the metal rather than to scratch the surface of the metal (Fig. 4-9).

In any case, as the electrode is brought close to the work, an arc will begin. As soon as you see a flash, you should raise the tip of the electrode up from the work until it is about 1/4 inch above the surface and the arc is still present. If the arc is held at this height in a steady position, a puddle of molten metal will begin to form where the arc touches the parent metal. At this time the electrode should be lowered toward the metal until it is at a height roughly

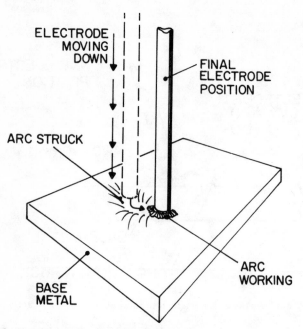

ELECTRODE MOVING DOWN

FINAL ELECTRODE POSITION

ARC STRUCK

ARC WORKING

BASE METAL

Fig. 4-9. Use the tapping method to strike an arc when using a dc arc welder.

equal to the thickness of the electrode. For example, if you are using a 1/8-inch diameter electrode, the tip of it should be about 1/8 inch above the work. This, incidentally, is about the time that beginners run into a freezing problem as mentioned earlier.

Striking a proper arc should be practiced until an arc can be struck, raised and then lowered into welding position without creating a short circuit and freezing the electrode. One very real indication that a proper arc has been struck is the arc will give off a crackling sound, very similar to bacon frying. This assumes, of course, that the welding machine has been set properly for current based on the electrode being used.

Cutting Metal

Arc welding equipment can also be used for cutting metal. As with arc welding, arc cutting uses a powerful electric force to melt the metal. The molten metal is then either oxidized or blown away with a special attachment which introduces compressed air. Generally speaking, special electrodes are used for arc cutting. These electrodes are coated with a special insulating material which does not conduct electricity. This coating allows the welder to touch the metal being cut without the possibility of freezing the

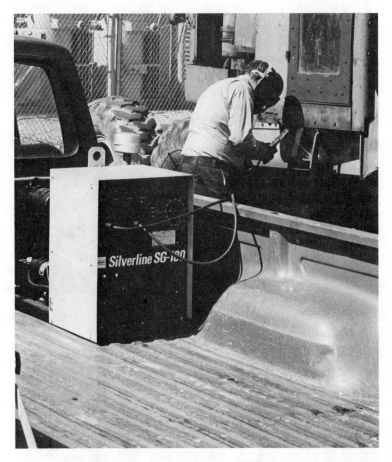

Fig. 4-10. It is also possible to cut with arc equipment.

electrode. The coatings on cutting electrodes act as an arc stabilizer, which concentrate the arc and at the same time intensify the arc's action. Cutting electrodes are nonconsumable, which means that they will retain their original diameters and lengths during cutting (Fig. 4-10).

Arc welding and cutting, as mentioned earlier, is not difficult to learn. It should not be attempted, though, without knowledgeable supervision.

TUNGSTEN INERT GAS WELDING

Tungsten inert gas welding, or simply TIG as it is called in the trade, was first introduced in the late 1940s and was an answer to welders' prayers. One of the problems when any two metals are joined with heat is that the molten metal is vulnerable to contamination from oxides in the atmosphere. The end result is a weld that may have weak spots. Before TIG welding was introduced, welders had to resort to various fluxes to remove contaminated material from the weld.

With the introduction of TIG welding, the problem of potentially weak joints and welds was eliminated. Basically, TIG welding is arc welding with the addition of a special inert gas, in most cases argon and/or helium, which shields the area being welded from airborne contaminated material. The gas itself is inert which means

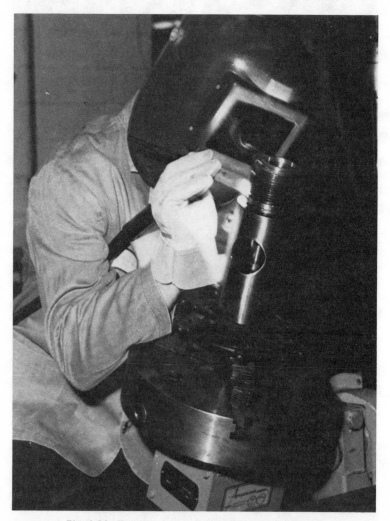

Fig. 4-11. Tungsten inert gas welding equipment in use.

that it has a neutral effect on molten metal. The end result is a very clean, strong weld.

Tungsten inert gas welding differs in another way from standard arc welding in that the electrode used to strike an arc is nonconsumable; it does not decrease in size or length. The heat from this tungsten electrode is sufficient to melt the edges of the metal being welded, but it is not hot enough to melt the tungsten. A welding rod, not unlike an oxyacetylene welding rod, is used as a filler material and is not attached to the electrode in any way.

Tungsten inert gas welding was originally developed specifically for welding manganese, aluminum and stainless steel. It is widely used to this day for nonferrous metals such as aluminum, nickel alloys, copper alloys and refractory metals. In the process of developing a new means of welding these metals used frequently in the aircraft industry, the inventors also developed a method that was fast, easy to learn and more economical. Tungsten inert gas welding also produces clean, beautiful welded joints (Fig. 4-11).

Unfortunately, the equipment necessary for TIG welding can be considered above the means of the casual, do-it-yourself welder. However, it is widely used in industry and specialty welding shops.

GAS METAL ARC WELDING

To say that the introduction of TIG welding opened the door to new welding methods is an understatement at the very least. Unfortunately, TIG welding has its limitations, the major one being that it is not suitable for welding metals with a thickness greater than 1/4 inch. *Gas metal arc welding* (MIG) was originally developed as a solution to welding metals thicker than 1/4 inch using the TIG principle of shielding the are being welded with an inert gas. Gas metal arc welding, when first introduced, was limited to rather thick welding applications. In a short period of time, certain refinements were introduced to the basic MIG system and made this type of welding into a very versatile means of joining metals.

Gas metal arc welding differs from TIG welding in one major way: the electrode is consumed during the welding process. Actually, the electrode in a MIG system is a continuous bare metal filler wire which is constantly fed through the handle of the unit. The wire makes continuous welding possible. This filler wire is fed into the arc and melts on its way to being deposited in the weld. As with TIG welding, MIG welding produces a strong, clean weld. Carbon dioxide is used in MIG welding as the shielding gas. Figure 4-12 shows an MIG welding unit in operation.

Fig. 4-12. Gas metal-arc welding equipment in use.

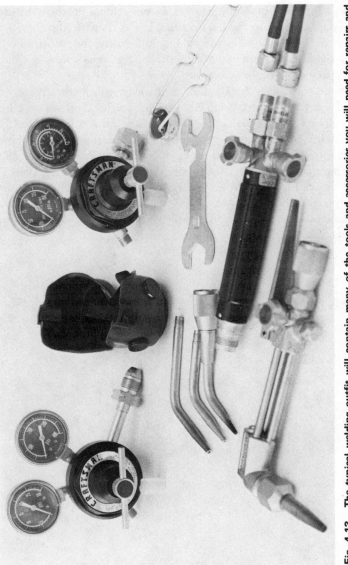

Fig. 4-13. The typical welding outfit will contain many of the tools and accessories you will need for repairs and building projects around your home.

Since the introduction of TIG welding in the 1940s and the later development of MIG welding, there have been newer additions to the welding field. American know-how knows no bounds when it comes to developing a new means for accomplishing a specific task. There are probably another dozen welding processes in use today in addition to the ones covered in this chapter. But these are all used for specific tasks for welding in industry and, as with both TIG and MIG, are not really designed for the home welder in either the price or application categories.

In all probability the do-it-yourself welder will first learn how to weld using a basic oxyacetylene welding outfit. This makes a lot of sense on several levels. The oxyacetylene process has the ability to handle a wide range of welding projects, certainly all of those encountered by the casual welder. Oxyacetylene welding and cutting is also a bit safer than other types like arc welding. There is less chance for injury to the user. Another strong point is that the equipment necessary to weld with gas is affordable. There are many welding outfits on the market which can be had (at this time) for under $200 (Fig. 4-13).

Chapter 5
Metals and Their Properties

Before the do-it-yourself welder can expect to achieve success in welding, he must have an understanding of the different properties of various metals. Since metals differ widely in strength, composition and melting points, it is impossible to think you can join any two pieces of metal simply by clamping them together and heating them up to the melting point until they join. Perhaps a very brief history of metals will help in a more thorough understanding of metals in general.

HISTORY OF METALS

Early man, by chance, discovered that certain earths, when exposed to extreme heat, left a residue after cooling that was very different from stone or the original earth. Man later discovered that this hard material could be pounded into various shapes which were useful for gathering food and making weapons.

The actual chemistry of metal was not really understood until fairly recently in man's evolution. Basically, metallic ores are oxides — compounds which are composed of metal and oxygen atoms. When these ores are heated with a fuel which gives off carbon monoxide, coke or charcoal, the ore will lose its oxygen as a result of the heat. The end product will be metal and slag. This process is called *reduction* by scientists. Because the ore atoms are rearranged, removing the oxygen, the end residue is metal.

Discovery of Copper

Copper was probably the first metal to be discovered. It is very common and can be heated out of ore with a relatively low temperature. Unfortunately pure copper is too soft to be used for weapons or tools. If *tin* is added to the molten copper, the cooled metal will be much harder than pure copper. The amount of tin added would, in fact, give the metal certain qualities. The addition of tin to pure copper produces an alloy called *bronze*. Man has known about bronze since about 2,000 B.C. This metal was first used for weapons, shields, bells and utensils.

If zinc is added to copper instead of tin, the resulting metal will be *brass*. Today there are a number of different brass compositions. Some of these contain iron and tin in addition to *zinc*. As man learned more about metal, he discovered that bronze and brass did not hold an edge for very long. Eventually tools and weapons were made from a stronger metal like *iron*.

Finding of Iron

Iron was first discovered as a residue left over from smelting other metals. Small bits and pieces of iron would be a common by-product in the slag of copper. Eventually some enterprising individual discovered that these pieces could be heated and then hammered into various shapes. Unfortunately, this early iron turned out to be not as strong as brass or bronze.

It is estimated by anthropologists that around 1,400 B.C. the Greeks discovered that heating and reheating iron, and then hammering it, would transform the iron into something far superior to bronze as far as strength and hardness were concerned. It was also discovered that repeated heating of iron would cause the metal to absorb carbon from the heating fuel. This excess carbon changes the iron into a carbon-iron alloy, steel, with a skin that is very hard. Today we do much the same thing to obtain case hardened steel, which is steel with a hard coating and a softer core.

As more experiments were conducted with iron it was discovered that if red hot metal was dropped into cold water, the result would be a metal that was harder than any other. Unfortunately, this also produced a metal that was very brittle when subjected to shock. The Romans, conquerors of the world, experimented a bit further and learned that much of the brittleness could be removed from this new metal if it was reheated and then quenched at a higher temperature. The metal retained its hardness but was no longer brittle, making it ideal for weapons. Archaeologists have discovered weapons with these properties dating back to about 1,300 B.C.

Steel Making Processes

It was almost 3,000 years (about 1720 A.D.) before the role of carbon was understood in the steel making process. A French chemist discovered that if certain amounts of carbon were added to molten iron, the resulting metal would be steel. The problem, at that time, was how much carbon should be added and how could the carbon be combined. A man by the name of Sir Henry Bessemer solved this problem when he invented a method of making steel which is still in use today. This discovery, in 1856, was called the *Bessemer process*. It uses a blast furnace to produce a low grade steel that is in great demand for various structural steel shapes.

The Bessemer process utilizes a controlled furnace that blows air through the molten iron. As this air escapes it takes burnt gases of carbon, manganese and silicon with it. The end result is a steel that is free of these elements. Then controlled amounts of carbon and/or other alloying elements are added to make a specific type of steel. Alloy steels, which we will cover later in this chapter, contain small amounts of tungsten, chromium, nickel, manganese and vanadium. Today steel and alloy steel are made in one of three processes: Bessemer, open-hearth and the electric arc furnace method (Fig. 5-1).

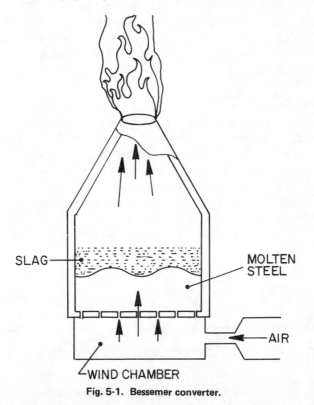

Fig. 5-1. Bessemer converter.

The do-it-yourself welder should be able to identify metals as closely as possible so he will be able to join these metals. Generally speaking, there are three basic areas of information, characteristics if you will, that the welder should be familiar with. These describe the metal and make it easier to identify. All metals have *mechanical*, *chemical* and *physical* properties.

PHYSICAL PROPERTIES OF METALS

Most of the physical properties of metals can be easily identified by sight. These properties include the *color* of metal and *magnetism*.

Color

Color as a physical property describes the surface appearance of the metal. Iron looks different from aluminum, gold is different from silver, copper looks different from zinc and so on. But there are also similarities. The visual test only serves to eliminate certain metals from consideration. Experienced welders and other people who have worked with metal for a long time can usually come pretty close to guessing what type of metal they are faced with. But the rest of us will have to rely on other methods for identifying metals. Nevertheless, the color test is valuable for grouping metals into possible categories.

Magnetism

All *ferromagnetic* metals — ferrous metals containing iron — have magnetic properties. In other words, a magnet will be attracted to these metals. Actually the magnetic test serves to group all metals into one of two rather broad categories, *ferrous* and *non-ferrous* metals. All ferrous metals are derived from iron and its alloys and are magnetic. Nonferrous metals are nonmagnetic and contain either no iron or insignificant amounts. Some examples of nonferrous metals are copper, bronze, brass, zinc and aluminum (Fig. 5-2).

Fig. 5-2. All ferrous metals are magnetic.

Melting Point

The *melting point* of a metal serves as probably a more accurate identification of specific metals and their alloys. Table 5-1 lists the various melting points of metals and alloys. One very good and accurate way to determine the melting point of a metal, aside from using a special thermometer, is to use a special crayon-like material which has a known melting point. These melt point indicators are sold at welding supply houses and are available in stick form (very much like a crayon), tablets and liquids. When

Table 5-1. Melting Points of Metals and Alloys.

CARBON	3500	
	3400	
	3300	
	3200	
CHROMIUM	3100	
PURE IRON	3000	WROUGHT IRON
MILD STEEL	2900	STAINLESS STEEL, 12%
	2800	CHROMIUM
	2700	COBALT
NICKEL	2600	SILICON
STAINLESS STEEL	2500	
19% CHROMIUM	2400	
MANGANESE	2300	
	2200	CAST IRON
	2100	
	2000	COPPER
	1900	
SILVER	1800	BRASS
	1700	
	1600	BRONZE
	1500	
	1400	
ALUMINUM	1300	MAGNESIUM
	1200	
	1100	
	1000	ALUMINUM ALLOYS
	900	MAGNESIUM ALLOYS
ZINC	800	
	700	
LEAD	600	
	500	
TIN	400	

Fig. 5-3. A temperature-indicating crayon.

these materials are heated, they will melt at a specific temperature — 2,800 degrees for example. When these special melt point indicators are used, the type of metal being considered can be fairly accurately determined (Fig. 5-3).

CHEMICAL PROPERTIES OF METALS

The chemical properties of metals include *corrosion, oxidation* and *reduction.* Corrosion is the chemical attack on metal by various elements in the atmosphere and can only happen on aluminum and its alloys. Oxidation in one form is rust and happens to all ferrous metals when certain oxides in or on the metal are combined with oxygen. Some metals, unless protected with a coating of some type, will combine with the oxygen in the atmosphere and produce rust. The ease with which some metals combine with oxygen can complicate the welding process since pure oxygen is used. In certain instances, when oxygen combines with the oxides in metals, reduction takes place. The volume of the metal will be reduced. Fluxes and shielding gases (as in TIG & MIG welding) are used to prevent reduction.

The chemical properties of metals can also be determined through certain chemical tests. For example, *chromemolybdenum*, a popular aircraft steel alloy, can be identified by immersing some filings from this metal in a dilute sulfuric acid solution. As these filings dissolve, the solution will turn deep green.

Chemical tests are not used with any frequency at the do-it-yourself level, but they are used often in industry to identify the composition of metals and their alloys.

MECHANICAL PROPERTIES OF METALS

Undoubtedly the most important characteristics of metals for the welder are the mechanical properties. This is a rather broad category

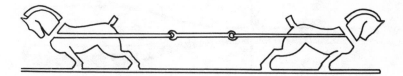

Fig. 5-4. Tensile strength of a metal indicates the amount of force necessary to pull that metal apart.

of descriptive terms used to identify the properties of metals. These include: *tensile strength, ductility, hardness, compressive strength, toughness, malleability, impact strength, fatigue, brittleness, stress, elasticity* and *creep.*   We will discuss each of these properties because of their importance to the modern do-it-yourself welder.

Tensile Strength, Ductility and Hardness

The tensile strength of a metal is the ability of that metal to resist being pulled apart. Imagine pulling a piece of taffy and you will have a good idea of tensile strength. The tensile strength of a metal is determined by placing a sample of a metal in a special machine, which is able to evenly pull with tremendous strength, and noting the point at which the specimen pulls apart. The tensile strength of a metal is expressed or measured in thousands of pounds per square inch (Fig. 5-4).

Ductility is a quality of metal that enables it to be formed or worked with force without breaking. When making wire, the ductility of the metal is important. Wire is made by pulling metal rod through a special die which forms the metal without breaking it (Fig. 5-5).

The term hardness implies solidity and firmness in a metal. Actually, hardness of a metal is the ability to resist forceable penetration of another material or metal. There are a number of methods used to determine the hardness of a metal. Probably the only test that can be done by the home welder is a simple file test. A

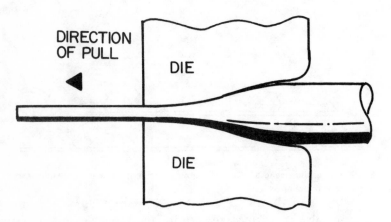

Fig. 5-5. A metal with good ductility can be worked with force without damage. Wire is made from metal with this quality.

standard machinist's hand file is used to file the metal. The results of the filing can then be compared with Table 5-2.

It should be noted, when performing the file test, that several areas of the metal should be tested and an average of the findings

Table 5-2. Guideline for Judging Hardness of Metals.

HARDNESS OF METAL (BRINELL METHOD)	WHEN FILE PASSED OVER THE SURFACE
100	Metal is very soft, easy to file.
200	Metal reasonably soft, slight pressure required on file.
300	Metal exhibits resistance but can be filed with pressure.
400	Metal is difficult to file, hard.
500	Metal appears almost as hard as file, very difficult to file.
600	Metal cannot be filed, damage to file teeth

determined.   Surface defects will, of course, have an effect on the results.   Therefore, extremely high or low readings should be disregarded.

Compressive Strength, Toughness and Malleability

The compressive strength of a metal is another mechanical property. A metal is said to have much compressive strength if it is able to withstand a force over a large area without deforming. It is important that the force be applied slowly and evenly rather than all at once. Striking with a hammer, for example, is not a good test for the compressive strength of a metal. One example of a metal that has good compressive strength is a gasoline engine head or block. This metal is able to withstand high compression during operation, but it will crack or break if dropped a few feet onto a hard surface.

A metal is said to be tough if it can withstand repeated applications and releases of force without change. An automobile leaf spring is one example of a tough metal. A good one will stand up under the force of a heavy load and return to its original position without change for many years (Fig. 5-6).

The malleability of a metal is that metal's ability to be worked cold without noticeable resistance. For example, low carbon steel can be hammered or bent into a shape but cast iron cannot. In this case the steel is malleable while the iron is not. Other examples are aluminum, copper, silver and gold which can all be cold rolled into thin sheets without deforming (Fig. 5-7).

Impact Strength, Fatigue and Brittleness

Impact strength of a metal is another metal property which is desirable in certain applications. A metal with good impact strength

Fig. 5-6. A metal is tough if it can withstand repeated force without changing. An automobile leaf spring must be made from tough metal.

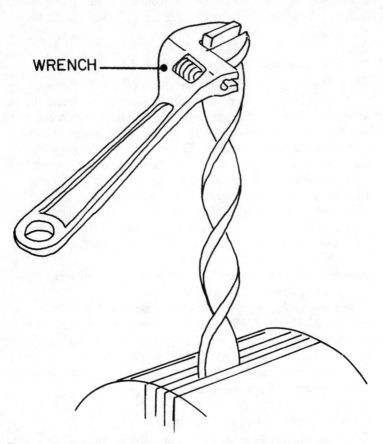

WRENCH

Fig. 5-7. Malleability is the ability of a metal to be worked easily without heat.

must be able to withstand repeated shock without fracturing or breaking down. An example of metals with high impact strength are those metals used for tools such as hammers and cold chisels (Fig. 5-8).

Metals are said to suffer fatigue when they do not hold up under continuous impact or load. Fatigue usually makes itself known in metal by a crack which lengthens in time. Metal fatigue usually happens when there are flaws in the metal combined with repeated applications of a heavy load. The flaw in the metal gives way due to the excess stress being applied (Fig. 5-9).

The brittleness of a metal indicates a lack of ductility. The metal will break, not bend. An example of brittleness is a pencil. It will not bend but will break when a certain amount of force is applied.

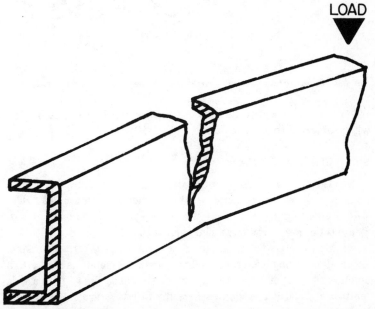

Fig. 5-8. A metal is said to have high impact strength if it can withstand repeated sharp force. Hammers and chisels are made from metals with high impact strength.

Fig. 5-9. Fatigue happens when a metal cannot withstand stress from a load.

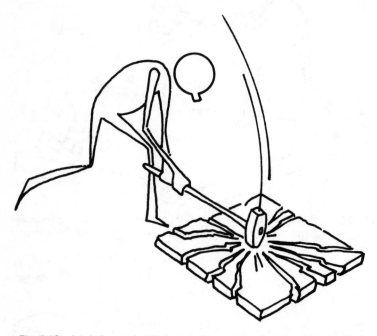

Fig. 5-10. A brittle metal will shatter when a heavy load is rapidly applied.

When a metal is said to be brittle, it will shatter like a piece of glass when subjected to a sudden shock (Fig. 5-10).

Stress, Elasticity and Creep

Stress refers to the amount of force, such as a load, applied against a metal surface. Some metals will bend when stress is applied. Heat will decrease the amount of stress necessary to deform metals. Strain is the result of stress when applied to metal.

Elasticity refers to the ability of a metal to return to its original shape after having been subjected to stress. All metals have a certain amount of elasticity, as well as having an elastic limit. The elastic limit of a metal is that point where it will not return to its original shape (Fig. 5-11).

Fig. 5-11. Metal which is elastic can be stretched to a certain limit while it is cold.

After the elastic limit of a metal has been reached, any additional force applied will cause a permanent change in the metal's form. This is known as creep.

METAL IDENTIFICATION

As you will see in later chapters, welding methods will differ according to the type of metal being repaired or welded. It is therefore important for the do-it-yourselfer to be able to identify metals and alloys of metals. This is particulary important when working with the various types of cast iron and different carbon content steels. As mentioned earlier, the color or general appearance of a metal can be an indication of the type of metal or, at the very least, the type of metal that it is not. See Table 5-3. It will be to the new welder's distinct advantage to obtain samples of as many different types of metals as possible so he will have some basis for comparison.

Table 5-3. Identification of Metals by Appearance.

	Alloy Steel	Copper	Brass and Bronze	Aluminum and Alloys	Monel Metal	Nickel	Lead
Fracture	medium gray	red color	red to yellow	white	light gray	almost white	white; crystalline
Unfinished Surface	Dark gray; relatively rough; rolling or forging lines may be noticeable	Various degrees of reddish brown to green due to oxides; smooth	Various shades of green, brown, or yellow due to oxides; smooth	Evidences of mold or rolls; very light gray	Smooth; dark gray	Smooth; dark gray	Smooth; velvety; white to gray
Newly Machined	Very smooth; bright gray	Bright copper red color dulls with time.	Red through to whitish yellow; very smooth	Smooth; very white.	Very smooth; light gray	Very smooth; white	Very smooth; white

	White Cast Iron	Gray Cast Iron	Malleable Iron	Wrought Iron	Low-Carbon Steel and Cast Steel	High-Carbon Steel
Fracture	Very fine silvery white silky crystalline formation;	dark gray	dark gray	bright gray	bright gray	very light gray
Unfinished Surface	Evidence of sand mold; dull gray	Evidence of sand mold; very dull gray	Evidence of sand mold; dull gray	Light gray; smooth	Dark gray; forging marks may be noticeable; cast-evidences of mold	Dark gray; rolling or forging lines may be noticeable
Newly Machined	Rarely machined	Fairly smooth; light gray	Smooth surface; light gray	Very smooth surface; light gray	Very smooth; bright gray	Very smooth; bright gray

Fig. 5-12. The chip test is one reasonably accurate means of identifying different metals.

Chip Test

Another way of identifying different metals is to remove a piece of the metal with a cold chisel and hammer (Fig. 5-12). Then carefully examine the piece. This is called the *chip test*. While there are obvious limitations, it is still a valuable test which offers fairly accurate results. See Table 5-4.

Spark Test

Still another valuable test to determine the type of metal in hand involves pushing a sample of the metal against a grinding wheel and watching the resulting sparks. As you might expect, this is called the *spark test*. It will give you a fairly accurate indication of what

Table 5-4. Identification of Metals by Chips.

	Copper	Brass and Bronze	Aluminum and Alloys	Monel Metal	Nickel	Lead
Appearance Of Chip	Smooth chips; saw edges where cut	Smooth chips; saw edges where cut	Smooth chips; saw edges where cut	Smooth edges	Smooth edges	Any shaped chip can be secured because of softness
Size Of Chip	Can be continuous if desired	Can be continuous if desired	Can be continuous if desired	Can be continuous if desired	Can be continuous if desired	Can be continuous if desired
Facility of Chipping	Very easily cut	Easily cut; more brittle than copper	Very easily cut	Chips easily	Chips easily	Chips so easily it can be cut with penknife

	White Cast Iron	Gray Cast Iron	Malleable Iron	Wrought Iron	Low-Carbon Steel And Cast Steel	High-Carbon Steel
Appearance Of Chip	Small broken fragments	Small, partially broken chips but possible to chip a fairly smooth groove	Chips do not break short as in cast iron	Smooth edges where cut	Smooth edges where cut	Fine grain fracture; edges lighter in color than low-carbon steel
Size of Chip		1/8 in.	1/4-3/8 in.	Can be continuous if desired	Can be continuous if desired	Can be continuous if desired
Facility Of Chipping	Brittleness prevents chipping a path with smooth sides	Not easy to chip because chips break off from base metal	Very tough, therefore harder to chip than cast iron	Soft and easily cut or chipped	Easily cut or chipped	Metal is usually very hard, but can be chipped

Courtesy of Union Carbide Company

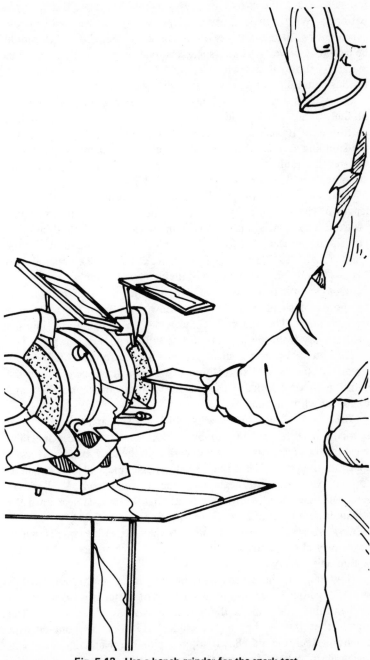

Fig. 5-13. Use a bench grinder for the spark test.

type of metal you have in your hand. The whole key to the spark test is to watch the sparks as various metals give off different spark patterns. With a little experience almost anyone, with a keen eye, can identify different types of metals and their alloys. It should be noted however that the spark test can only be used for ferrous metals — those metals which contain iron in some form — as non-ferrous metals do not have any appreciable spark.

Since the spark test is neither difficult or beyond the capabilities of the average home welder, it will be helpful to explain how it is performed. Basically, a piece of metal is lightly touched against the spinning surface of a grinding wheel. The stream of sparks that will result should then be observed against a dark background and at an angle which permits a good view. Needless to say, some type of clear eye protection should be worn while doing this test. The length of the stream of sparks will be directly related to how much pressure is exerted against the grinding wheel. The three things that you should be watchful for are the color, volume and nature of the sparks (Fig. 5-13). The stream of sparks can then be compared with Fig. 5-14 to determine the type of metal being ground.

Wrought iron is iron with no carbon. It has a spark stream that is composed of small particles which flow away from the wheel in a straight line. The sparks become wider and brighter as they get further away from the wheel until finally they go out.

Mild steel contains a small percentage of carbon. It will have a stream of sparks similar to those of wrought iron, except with the addition of tiny forks in the sparks. The carbon causes the metal to appear white in color.

Tool steel will produce sparks which are similar to the mild steel sparks, but with a greater number of forks in the stream. Generally speaking the greater the amount of carbon in the steel, the more forked sparks in the trail. High carbon steel, for example, will have a blaze of forked sparks that will begin very close to the wheel. Mild carbon steel will have sparks with few forks that will not appear until the stream is about 1 foot away from the grinding wheel.

Cast iron will have a stream of small repeating sparks which begin very close to the grinding wheel and do not last. The spark stream will be very short.

Now that we have discussed the properties of different metals as well as how to determine what type of metal is at hand, it will be helpful to talk about the actual metals and their alloys. The metals that the average home welder will come in contact with are iron, steel and its alloys, copper, aluminum and its alloys and magnesium. A brief description of these metals follows.

Fig. 5-14. Metal salvage yards usually carry a large selection of steel and other metals useful to the home welder.

IRON

As explained earlier in this chapter, iron has been in use for quite a long period of time. Iron is extracted from iron ore and is the second most common metal on the earth. Aluminum is the most common. Iron is found only in combination with other elements, which must be removed to produce iron. It is widely used commercially for such things as bridge parts and castings. Wrought iron was the type of iron used before steel could be produced economically. Wrought iron is very low in carbon as a result of the way it is made.

Cast iron is wrought iron with the addition of at least 1.7 percent carbon and certain quantities of sulfur, phosphorus, manganese and silicon. Gray cast iron is cast iron with the addition of *graphite*. Graphite is a form of free carbon that combines with pure iron during smelting. Its addition produces a form of iron which is softer than cast iron.

White cast iron is produced when scrap steel and pig iron are melted in an air furnace or *cupola*. This special process produces a white cast iron which is very hard and brittle. Generally white cast iron is produced along the way to making malleable cast iron.

When white cast iron is annealed, the end product will be malleable cast iron. *Annealing*, basically, is a special heating process followed by a very slow cooling that creates an iron which is malleable. Malleable cast iron is a result of a special process which heats up castings (made from white cast iron) and keeps them at a specific controlled temperature. Then the castings are cooled over a period of about 24 hours. Although the finished product is not by definition malleable, it possesses great toughness and is not brittle.

STEEL

As you know from the discussion earlier, steel is an alloy of iron with a carbon content of no more than 1.7 percent. Steel is easily the most common metal for the welder, and there are many different types. By varying the content of carbon, or by adding other elements, different types of steel can be produced. Steel with a low carbon content can be soft and easily worked by hand with light tools; steel with a high carbon content will be very hard and may have other characteristics as well.

There is a very broad range of carbon steel in use today. Steel can be classified with either a number, such as 440 carbon steel, or simply by designations such as machine steel or tool steel. In any case these designations will indicate the various qualities of a particular steel. For example, machining steel has a low carbon content,

0.10 percent to 0.20 percent, and is easily worked by a machine such as a lathe. Tool steels have a higher carbon content, at least 0.60 percent but lose their hardness at high temperatures because the carbon leaves the metal when heat is applied.

Carbon can be added to molten steel up to 0.83 percent. After this point, the tensile strength of the steel will be reduced. The higher the carbon content in the steel, the greater is the steel's resistance to wear. There is also a sacrifice as far as strength and toughness are concerned.

Alloy steels are created when other elements are added to plain carbon steel. Basically an alloy is the combination of any two metals in any proportions. In the case of steel, there are many different elements and metals which can be added to produce a new steel with certain qualities. See Table 5-5 for alloying elements of steel.

COPPER

Having been in use for over 4,000 years, copper has many applications around the home and in industry. Probably the greatest use is in plumbing pipes and water lines. Copper is very resistant to corrosion and oxidation. One of the unique properties of copper is that when it cools from a molten state, it expands. This is a favorable reaction when welding pipes and joints. When copper is heated to just below its melting point, (about 2,000 degrees Fahrenheit) it becomes as brittle as glass. Copper is a tough (tensile strength about 20,000 psi) ductile metal which can be easily worked cold or with heat.

ALUMINUM

This metal is widely used in industry as well as around the home. Pure aluminum has a tensile strength of about 13,000 psi which can be increased with alloying and heat treatment to about 80,000 psi. Aluminum can be worked by any method used commercially. Because aluminum is half the weight of iron, it is used extensively in the aircraft industry.

The surface of aluminum is oxidized when it comes in contact with the atmosphere, which can have an effect on welding. This semi-transparent oxide film must be removed before welding can be expected to be successful. Aluminum poses special problems for the welder which are covered later in this book.

MAGNESIUM

Magnesium is the lightest metal in use today. It equals about one-fifth the weight of copper and less than two-thirds the weight of

Table 5-5. Principal Alloying Elements of Steel.

Element	Melting Point (°C)	Application	Reason for Use
Aluminum	658	Little aluminum remains in steel.	Deoxidizes and refines grain. Removes impurities.
Chromium	1615	Stainless steels, tools, and machine parts.	Improves hardness of the steel in small amounts.
Cobalt	1467	High speed cutting tools.	Adds to cutting property of steel, especially at high temperatures.
Copper	1082	Sheet and plate materials.	Retards rust.
Lead	327	Machinery parts.	Lead and added tin form a rust resistant coating on steels.
Manganese	1245	Bucket teeth. Rails and switches.	Prevents hot shortness by combining with sulfur. Deoxidizes. Increases toughness and abrasion-resistance.
Molybdenum	2535	Machinery parts and tools.	Increases ductility, strength and shock resistance.
Nickel	1452	Stainless steels. Acid-resistant tools and machinery parts.	In large amounts — resists heat, adds strength, toughness, and stiffness to steel.

Table 5-5. Principal Alloying Elements of Steel (cont.).

Element	Melting Point (°C)	Application	Reason for Use
Phosphorus (provided by ore)	43	Some low-alloy steels.	Up to 0.05 percent increases yield strength.
Silicon	1420	Precision castings.	Removes the gases from steel. Adds strength.
Sulphur	120	Some machined pieces.	Adds to the steel's machinability.
Tin	232	Cans and pans.	Forms a coating on the steel for corrosion-resistance.
Titanium	1800	Used in low-alloy steels.	Cleans and forms carbide.
Tungsten	3400	For magnets and high-speed cutting tools.	Helps steel retain hardness and toughness at high temperatures.
Vanadium	1780	Springs, tools, and machine parts.	Helps to increase strength and ductility.
Zinc	420	Wire, pails and roofing.	Forms a corrosion-resistant coating on steel.
Zirconium	1850	Machine parts and tools.	Deoxidizes, removing oxygen and nitrogen. Creates a fine grain.

aluminum. Although magnesium can be worked by all methods, it is subject to rapid oxidation. Therefore, special welding methods must be used. Magnesium is a metal that combines lightness in weight with excellent strength. Magnesium alloys can have a tensile strength of up to about 50,000 psi.

SOURCES OF METALS

Before you can begin to learn how to braze, cut or weld, you must lay your hands on a supply of metal. Probably the best place to start your search is in the yellow pages of the telephone directory. The categories to look under are steel, steel fabricators, steel products, metal, salvage, sheet metal and metal stamping. You can also look under specific metal headings such as brass, iron, aluminum and stainless steel, just to name a few. You might also check under steel mills, if you know of any in your geographical area. Metal scrap yards are also a good source of relatively inexpensive steel and other types of metal. Your welding supply house, where you buy your oxygen and acetylene, should be able to tell you where you can purchase steel for welding projects. Local welders are another good source of information about where to buy metal. In many cases, these people who make a living welding and repairing metal will often have scrap metal in limited quantities for sale. Professional welders are also a very good source of technical information and advice. It can be to your advantage to establish a relationship with a few.

Obtaining New Steel

In the beginning you should work with new steel. There will be fewer problems to deal with than if you were to begin working on rusted scrap steel, for example. This will mean a trip to a local steel fabricator. Here you should find a large selection of steel including sheets, plate, tubing, angle, rod and strap steel. These companies operate by purchasing large quantities from steel mills and keeping inventories for steel workers who most commonly buy in small amounts (Fig. 5-15).

It has always been my experience that the people who operate steel fabrication firms are quite cooperative. Most large companies will often be very helpful in finding the specific metal and gauge you need. If you are looking for a specific type of metal that is not generally stocked, certain sheet steel, for example, a steel fabricator will be the place to go. Keep in mind that special orders usually require a deposit.

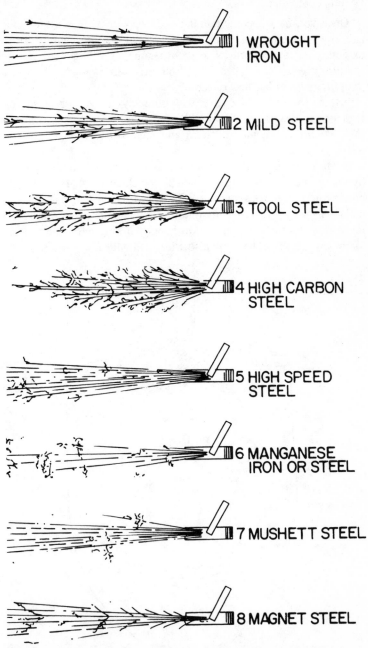

Fig. 5-15. Representations of the sparks obtained when various metals are lightly pressed against a grinding wheel.

Getting Scrap Metal

Once you have become familiar with the various brazing, welding and cutting techniques using new steel, you will probably want to extend your basic knowledge to other metals — high carbon and stainless steels — and bronze. You may also want to try metal sculpture with either new or scrap metal.

Again, the steel fabricator will probably be your best source. It is a wise idea to obtain samples of many different types of metals such as chromium, molybdenum and high carbon steel so you can widen your experience.

Scrap metal is usually very reasonably priced and commonly sold by the pound. Many interesting shapes, thicknesses and types of both nonferrous and ferrous metals can be had with a single trip to the local metal salvage yard. Keep in mind that it will be necessary to devote more time to surface preparation when working with scrap metal. Many sculpters feel the extra effort is worthwhile (Fig. 5-16).

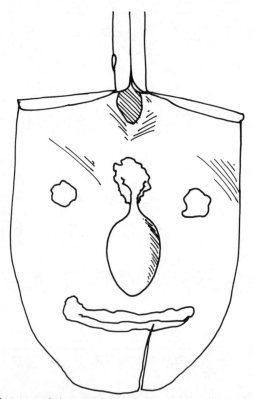

Fig. 5-16. Scrap metal that you may have laying around your welding shop can often be turned into an interesting shape or design.

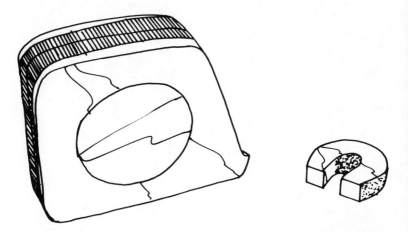

Fig. 5-17. A ruler and a magnet are handy tools to bring along to the steel yard.

When you pay a visit to your local steel fabricator or metal salvage yard, you should be prepared for the possibility of getting a little dirty from rummaging through scrap piles. Old clothes, boots, leather gloves and a hat will do much to protect you from the generally dirty conditions. A small pocket ruler and a magnet are useful tools to take along as well (Fig. 5-17). You should also be prepared to carry the metal home. Undoubtedly the best way is in the back of a pickup truck. If you don't own one, ask a friend to go along with you. As a last resort you can pile the metal on top or inside your automobile. Roof racks and rope will save a bit of wear and tear on the interior.

When you get the steel home, you will want to store it so it won't rust. If you don't have room inside, you will obviously have to store the metal outdoors. At the very least you should cover the steel with a plastic or canvas tarp.

Chapter 6
Welding Supplies

Your local welding supply house is your best source for welding supplies and technical information. If you ever have a question about a specific welding project, this is the place to ask. Many of these supply houses also offer instruction and other services (Fig. 6-1). Your welding supply house will be your source for oxygen and acetylene. You should also find a wide selection of tools, accessories, fluxes, and welding, brazing and special purpose rods. This chapter will focus on rods and fluxes.

WELDING RODS

During the welding process it is often necessary to add more metal to the molten area to make the weld stronger and to fill in the area. This is accomplished through the use of a special *welding rod*. It is very important that the welding rod being used matches as closely as possible the metal being welded, the parent metal, for a strong fusion of the metal. During welding the rod melts and deposits its metal into the molten parent metal and contributes to the strength of the weld. If the welder were to choose the wrong type of welding rod, the end result may be a weak and ineffective joint.

As you might expect, welding rods are available in a wide variety of sizes and compositions. There is easily a welding rod for every type of parent metal. The standard sizes of welding rods are 36 inches long for all types except cast iron which is sold in 24-inch lengths. Diameters of gas welding rods range from 1/16 to 3/8-inch (Fig. 6-2).

The most commonly used gas welding rods are mild steel, alloy steel and cast iron. Mild steel welding rods are most commonly covered with a copper coating which protects the rod from rust during storage. Generally speaking, mild steel welding rods are low in price and have a wide application for welding all types of mild

Fig. 6-1. A metal salvage company is a very good place to buy metal for projects.

Fig. 6-2. An assortment of welding and brazing rods. The square rod is for cast iron welding.

steel with a low carbon content. The tensile strength of mild steel welding rods is around 52,000 psi.

Alloy steel welding rods (for oxyacetylene welding) are used for relatively low carbon alloy steels. These rods are general purpose welding rods which are frequently used for pressure systems such as pipelines. The tensile strength of these rods is about 62,000 psi.

Cast iron welding rods are available in several different types, depending of course on the application. For example, cast iron rods are available for welding gray cast iron, molybdenum cast

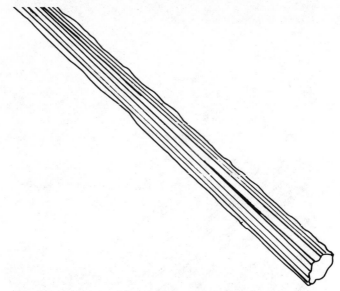

Fig. 6-3. Cast iron welding rods are square rather than round.

iron (an alloy of cast iron) and straight cast iron. As mentioned, cast iron rods are available in 24-inch lengths. They are also most commonly square rather than round (Fig. 6-3).

In addition to the three basic types of welding rods, there are many other special purpose rods. These rods are used when joining other types of metals such as stainless steel and aluminum. Inquire at your local welding supply house for information about the various types of welding rods. You might also want to pick up an assortment of welding rods for experiments.

BRAZING RODS

Brazing rods are used when two metals will be joined by the brazing method. Basically, brazing rods are either bare or flux-coated bronze or brass rods which are used for brazing or braze welding steel, cast iron, brass and bronze. Standard bronze brazing rods have a tensile strength (when used on steel) of about 50,000 psi. The melting point of standard bronze brazing rods is much less than that of other metals, around 1,600 degrees Fahrenheit.

In addition to the popular bronze brazing rods, other types are also available for special metal joining situations. These special rods include those made from manganese/bronze, deoxidized copper and nickel silver brazing rods. Brazing and all types of rods for this metal joining process are covered in detail in Chapter 8.

SURFACING RODS

Still another type of rod used by the welder is a hard *surfacing rod*. Briefly, certain tools and machinery which are subjected to abrasion, such as a plow blade or the blade on a bulldozer, should be protected against wear by hardening or coating the edge of the tool that receives the punishment. One way to do this is to heat the forward area with a welding torch and then melt a special metal over it (Fig. 6-4). This additional new material will be extremely resistant to wear and will help prolong the life of the tool. Generally speaking, the most common hard surfacing materials are nonferrous cobalt-chromium-tungsten alloys. These include nickel base, iron base and tungsten carbide materials.

Fig. 6-4. Application of a melted metal to the underside of a plowshare.

Fig. 6-5. Fluxes aid in joining certain types of metals and are often required for soldering and brazing.

FLUXES

Successful metal joining, brazing or welding, is directly related to how clean the metals are prior to joining. In some metals such as aluminum, the formation of oxides during the joining process presents a special problem. These oxides must be removed from the surface being welded or brazed to insure a solid bond. In Chapter 4 we briefly discussed shielded arc welding (TIG and MIG welding) and how an inert gas is sprayed over an area during the welding process to prevent contamination. During oxyacetylene welding, there is no such protection, so another means must be used. This protection is accomplished through the use of special fluxes (Fig. 6-5).

A flux, according to the American Welding Society, is a material used to prevent, dissolve or facilitate the removal of oxides and other undesirable substances from the area being brazed or welded. All fluxes are nonmetallic and fusible. Fluxes cause a chemical reaction to take place resulting in a slag that floats to the top of molten metal during welding or brazing. As you may have guessed, in order for this chemical reaction to take place, there are fluxes

Fig. 6-6. Special fluxes are almost always required when brazing.

for different types of applications. When welding some types of steel, flux is generally not required because the melting point of the steel is about the same as that of the oxides. Other metals, such as aluminum and iron, do not possess these qualities. Therefore, a flux must be used to lower the melting point of certain oxides so they can be removed and a strong joint will result. Special fluxes are always required when brazing metal (Fig. 6-6).

Fluxes are available for brazing, welding and soldering. They are sold as powder, paste and liquid. Flux-coated brazing and welding rods are also available for special applications. It is always a good idea to check at your local welding supply house whenever you are in doubt as to the right type of flux for a particular project.

Chapter 7
Soldering

There are three ways of joining metal: soldering, brazing and welding. Of these, the strongest, of course, is welding wherein the edges of the metal are heated up to the melting point. Then they fuse together Brazing is similar except that the metals are not brought up to the molten state, but a metal filler material is. Soldering metals produces the weakest joint of all three methods. In many cases, such as joining plumbing connections, strength is not as important as a complete sealing of the surfaces (Fig. 7-1).

Fig. 7-1. Soldered joints are never very strong, but in some cases strength is less important than a good seal.

Table 7-1. Temperature Ranges for Metal Joining Processes.

	2,800*
WELDING	
	2,000
BRAZING	
	800
SOLDERING	
	0

*Degrees Fahrenheit

The principle of metal *soldering* is really quite simple. The pieces are joined without bringing either up to the molten stage. Instead, a special metal, solder, is used as a bridge between the two pieces. Where soldering can be used, it is quite effective. Technically the process is called *adhesion*. In fact, what takes place is the molecules of solder mingle with the parent metal's molecules and form a strong bond. In some cases the ingredients in the solder, nonferrous metals, form a surface alloy on one or both of the metals being joined. This forms a solid bond between the metals.

Soldering is similar to brazing in that a nonferrous metal is used to bridge the space between the two metals. The major difference between these two methods, however, is the temperature at which the joining nonferrous metal melts. All soldering is done at temperature below 800 degrees Fahrenheit, while all brazing is done at temperatures above 800 degrees Fahrenheit. To give you an idea of the temperatures at which soldering, brazing and welding are accomplished, please refer to Table 7-1.

TOOLS

Since not as much heat is needed for the soldering process, different tools are used than for brazing or welding. Soldering can be accomplished with any number of low heat generating tools such as a soldering iron, soldering gun, or any of the gas (acetylene, propane, natural gas, Mapp gas) torches specifically designed for the purpose. It will be helpful to discuss each of these soldering tools and explain their possible uses.

Soldering Irons

Soldering irons are probably the simplest type of soldering tool. A typical soldering iron is simply a piece of steel rod with a wooden handle on one end and a large copper tip on the other end. Sizes can range from the small pencil-like lightweight types for small soldering

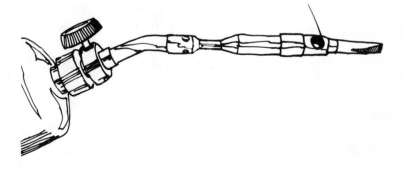

Fig. 7-2. A typical soldering iron consists of a wooden handle, steel shaft and a tip.

projects to large two-handed units used for special industrial applications. This type of soldering iron must be heated up to a temperature hot enough to melt the solder, up to 800 degrees Fahrenheit (Fig. 7-2). This is generally accomplished through the use of a special furnace which is commonly gas operated, although some furnaces are electrically heated (Fig. 7-3).

There are other types of soldering irons which are much more portable because they are heated internally, either electrically or with some type of bottled fuel. Both of these types of soldering

Fig. 7-3. Old soldering irons, since they had no power of their own, had to be heated in a special furnace which was most commonly gas-fired.

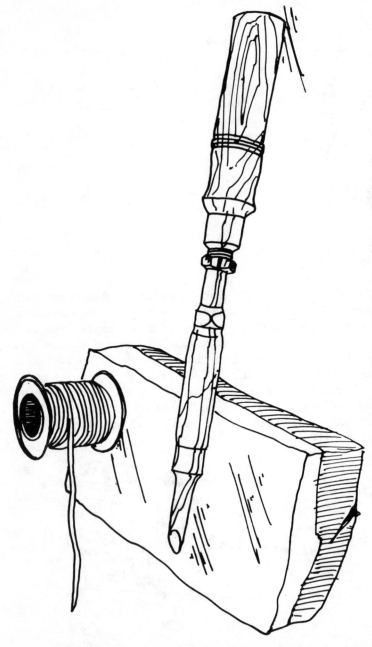

Fig. 7-4. This soldering iron tip is heated by the flame from a propane torch.

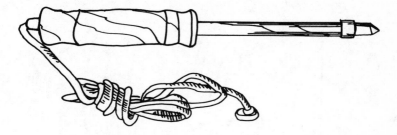

Fig. 7-5. An electrically heated soldering iron is very handy for continuous soldering.

irons are illustrated in Figs. 7-4 and 7-5. The electrically heated soldering iron is best suited for continuous delicate work, while the gas powered unit has its limitations because of the resultant flame below the tip.

Soldering Guns

Soldering *guns* are quite popular with do-it-yourselfers. They are fast, quick heating and quite suitable for small repairs and electronic building projects. There are a number of these soldering guns on the market, usually in some form of soldering kit which contains an assortment of tips for different types of soldering (Figs. 7-6 and 7-7).

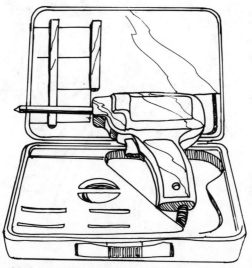

Fig. 7-6. A soldering gun kit contains several different size tips and is handy for a wide range of soldering projects.

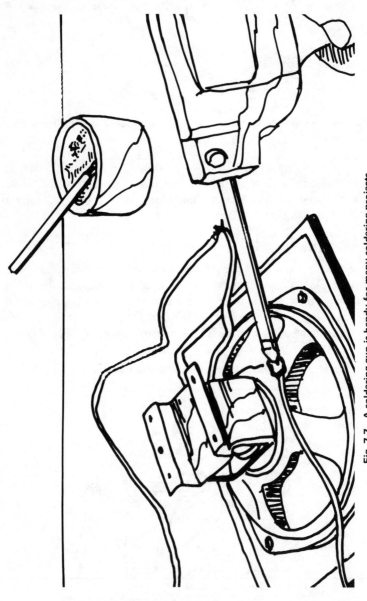

Fig. 7-7. A soldering gun is handy for many soldering projects.

Fig. 7-8. A propane torch kit has many uses around the home and workshop.

Torches

Still another tool used for soldering work is a gas-powered, hand-held *torch*.   There are several variations of this type depending on the intended application.   The standard do-it-yourselfer model has many uses around the home (Fig. 7-8).   The flame itself is used for heating the metal and melting the solder.   Needless to say, a flame-powered torch can only be used for certain types of soldering (Fig. 7-9).   It is a very handy tool for soldering the joints in a copper pipe plumbing system. The torch can also be used for thawing frozen water lines, removing paint and even heating vinyl asbestos floor tile prior to installation.   The most common fuel for this type of torch is *propane*, sold in pressurized cylinders which contain enough propane to run the torch for about two hours. Presently available are cylinders filled with *Mapp gas*.   This gas gives a hotter flame than conventional propane.

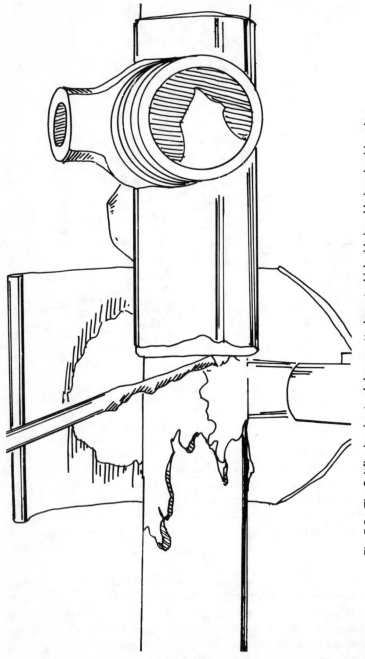

Fig. 7-9. The Spitfire brazing torch has an attachment which aids in soldering plumbing pipe.

These types of torches are commonly sold in a torch kit which contains various attachments such as a soldering tip and various size burner heads (Fig. 7-10).

Professionals who must solder many joints in the course of a day's work will usually have a special torch for this purpose. These torches, popular with plumbers, have a hand-held torch unit connected by a pressure hose to a regulator which fits into a valve on top of a 25-pound tank (or larger) of acetylene. The torches can also run on LP gas and are a work saving tool. This type of low heat torch outfit is standard equipment at almost any radiator repair shop (Fig. 7-11).

Copper Tips

All soldering tools with *copper tips* — excluding hand-held torches without tips — should provide years of dependable service with little care. Undoubtedly the most important part of all soldering tools is the tip of the unit. This part is the only one which comes in contact with the work. In order to insure that this important part is able to do its intended task of heating the metal and melting the solder, there are a few things which you must do to it. We will discuss these maintenance procedures in detail.

The tip of a soldering tool should never be allowed to overheat. On electrically powered units this is rarely a problem as the electrical current is controlled. But soldering irons which must be heated in a furnace are often prone to overheating. Overheating of the tip will result in excessive scaling of the soldering iron core and carbonizing of rosin fluxes. It will be difficult to properly tin the overheated tip. One indication of overheating of the soldering iron tip is a greenish tinge that will appear when the tip is hotter than it should be. Most authorities recommend that a soldering iron tip never be brought up to a temperature in excess of 750 degrees Fahrenheit. In most cases, soldering can be accomplished at lower temperatures anyway, so there is little point in bringing the iron up to this temperature.

Fluxes, which we will discuss in more detail later, can have a corrosive effect on soldering tips. While fluxes are necessary, you should use the mildest flux possible for the particular job at hand. When you have completed a soldering project, you should clean the tip of flux residue with a damp cloth or steel wool.

If your soldering tool is electrically operated, you should unplug it when not in use. This is particularly important for those soldering tools which remain hot as long as they are plugged in. For soldering

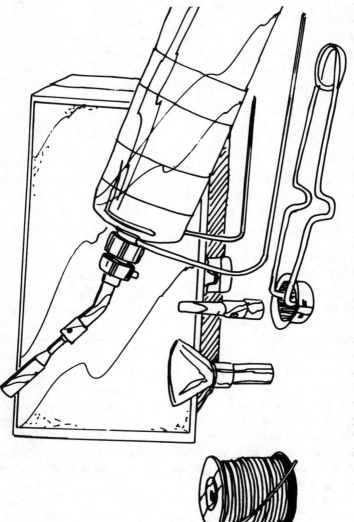

Fig. 7-10. A good propane torch kit will contain several useful attachments.

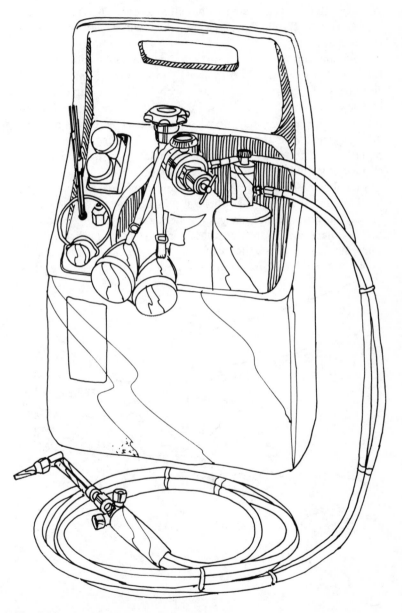

Fig. 7-11. A portable torch outfit is suitable for light cutting, welding and soldering.

guns, with a trigger that activates the heat source, this is not quite as important.

While using the soldering tool, you will achieve better results if you keep the tip well coated with molten solder. This is commonly called "tinning the tip," a procedure which will be covered later in this chapter.

Soldering tips will become dirty in time through normal use. For this reason all tips should be cleaned periodically. In most cases a cleaning with a wire brush or steel wool will restore the tip to a like-new condition. Care must be exercised when doing this, especially in the case of copper-coated tips, as the coating can be worn off if cleaned with too abrasive a tool. A grinding wheel should never be used for cleaning a tip, for example. After the tip has been cleaned, it should be immediately retinned (Fig. 7-12).

Probably more damage is done to soldering tools through rough handling and dropping than through normal wear and tear. It will be to your advantage to use your soldering tool properly, according to the directions of the manufacturer. Be careful when using a soldering tool because the possibility of burning yourself always exists. A soldering iron holder is some good insurance against

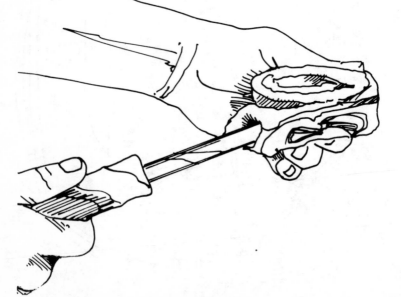

Fig. 7-12. Soldering iron tips should be cleaned with steel wool before each use. This will make soldering both easier and quicker.

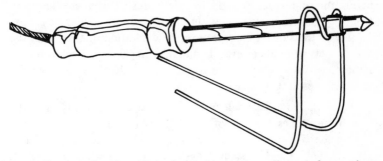

Fig. 7-13. A serviceable soldering iron holder can be easily made from a piece of welding rod or coat hanger.

accidental burns and also serves to keep the tool in its proper place. You can usually fashion a soldering tool holder out of a piece of steel welding rod or even a wire coat hanger (Fig. 7-13). You can also rest the hot iron on a brick or other suitable surface.

CHOOSING THE PROPER SOLDER

The key to successful soldering lies in choosing the proper solder for the job at hand. As you might expect, there is quite a large selection of different types of solders currently being sold. In truth, the two most common types of solder used outside of industry are those designated as 50-50 and 60-40. These two are the easiest for the do-it-yourselfers to use. Type 50-50 solder is 50 percent tin and 50 percent lead, while type 60-40 is 60 percent tin and 40 percent lead. The percentage of tin is always listed first. A chart of the six most commonly used solders is shown in Table 7-2. All solders except the high temperature, 5-95 solder, melt at a rather low temperature. You will also notice, however, that the more lead in the solder, the higher the temperature must be

Table 7-2. Common-Tin-Lead Solders.

ASTM NUMBER	% TIN	% LEAD	MELTING TEMPERATURE	FLOWING TEMPERATURE
5 A	5	95	572	596
20 A	20	80	361	535
30 A	30	70	361	491
40 A	40	60	361	455
50 A	50	50	361	421
60 A	60	40	361	374

*Degrees Fahrenheit

before the solder will flow or become liquid. This is due largely to the fact that the greater the amount of tin in a nonferrous alloy, the lower the flowing temperature. The addition of tin lowers the alloy's melting temperature. In an effort to give you a better understanding of these six common solders I would like to briefly explain the uses of each (Fig. 7-14).

- **5A Solder.** This solder has better strength quality than the other types because of the high lead content. It is a good solder for torch soldering where a flame is used rather than a heated tip. The solder can be difficult to work with unless you are quite experienced. For that reason, it is generally a poor choice for the do-it-yourselfer. Nevertheless, 5A solder has the ability of providing a very strong joint.

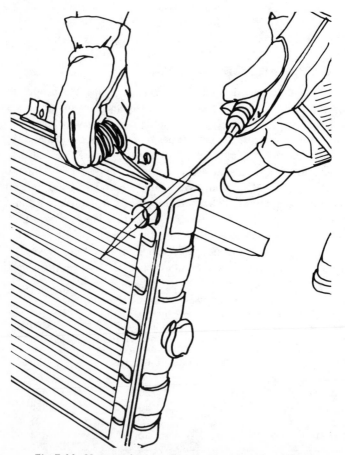

Fig. 7-14. Most repairs on radiators are made with solder.

- **20A Solder.** This type is used extensively in certain industrial soldering tasks such as sealing automobile radiators and auto body work (Fig. 7-15). This is a good solder to use where the finished joint or joints will not be exposed to high temperatures which can cause the joint to deteriorate.

- **30A Solder.** It is similar to 20A solder in that it is widely used for industrial work. This is a common solder for automobile body repair and torch soldering of all types.

- **40A and 50A Solder.** These are probably the most widely used general purpose solders available today. Both types are used for plumbing pipe joints, roofing seams (where tin or

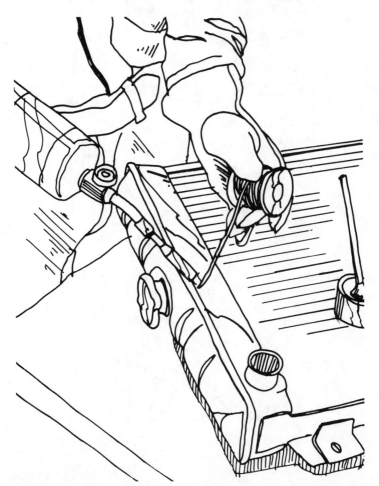

Fig. 7-15. One of the best solders for repairing an automobile radiator is 20A.

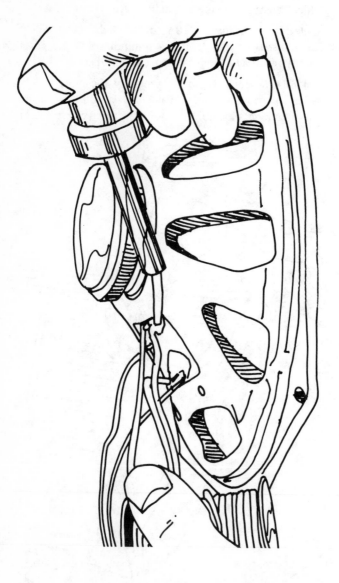

Fig. 7-16. General purpose solders include 40A and 50A.

copper is the covering material) and for electrical circuit connections such as radio and television. When used for electrical work, the solder usually has a rosin core instead of being solid. Rosin core simply means that the solder contains a noncorrosive flux which comes from pine trees (Fig. 7-16).

- **60A Solder.** This is often called fine solder and is used wherever temperature requirements are critical, such as in precision instruments.

Another group of solders worth mentioning is called the silver solders. This group can contain either lead-silver, tin-silver or a combination of all three nonferrous metals. Silver solders are used for special joining tasks, such as for certain types of metal or where electrical conductivity is important, particularly radio or other electrical component circuitry. Silver soldering produces a strong joint. Generally higher soldering temperatures and special fluxing techniques are called for, not to mention greater skill with the soldering tool.

FLUXES

One of the requirements of successful soldering is that the metal surfaces must be chemically cleaned. They must remain in this state until the solder has hardened (Fig. 7-17). Flux was designed

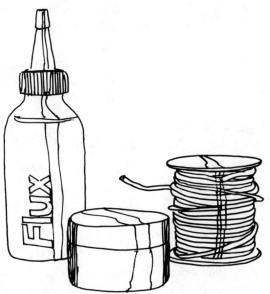

Fig. 7-17. Fluxes and solders go hand in hand.

for this purpose. When properly used, it not only cleans the surfaces but seals them for the atmosphere during the soldering. Additionally, some fluxes add alloying elements to the solder or parent metal while other fluxes reduce the surface tension on the metal, which increases the fluidity and adhesion of the solder.

At this point in time, there is no universal flux. Instead, there are several different types of fluxes designed for certain types of metals. Fluxes are available in liquid, powder, paste and solid form.

Noncorrosive

All fluxes can be grouped in either the noncorrosive or corrosive category. Generally speaking, noncorrosive fluxes are mild or weak in their fluxing action. These types are recommended for all electronic and other critical soldering projects. Noncorrosive fluxes are generally suitable for soldering clean copper, brass, bronze, tinplate, cadmium, nickel and silver. Noncorrosive fluxes leave an unsightly brown stain on the surface. Since this residue is not electrically conductive, it must be removed after soldering.

Corrosive

Corrosive fluxes are composed of inorganic ingredients. These fluxes are used when a rapid highly activated fluxing action is desirable, such as when soldering cast iron, sheet metal, and copper or brass. Corrosive fluxes generally work well with any of the methods commonly used for soldering.

Fig. 7-18. Acid flux.

Corrosive fluxes leave a residue after soldering that can attack chemically the joint just soldered as well as surrounding metal areas. For this reason, the residue from these fluxes must be removed after soldering. Corrosive fluxes are unsuitable for certain types of soldering projects such as sealed containers and electrical connections because of the corrosive action of the flux. One of the advantages of corrosive fluxes is that they remain relatively stable over a wide range of temperatures, which makes them suitable for the soldering projects requiring higher temperatures.

Probably the most common type of corrosive flux is called "acid" flux (Fig. 7-18). A common formula for this flux consists of 70 percent zinc chloride and 30 percent ammonium chloride. This type is called acid flux because as the flux is heated and melts, it forms hydrochloric acid. The acid dissolves the oxides and permits the solder metal to adhere to the parent metal. As mentioned, corrosive flux residue must be removed after the soldered area has hardened (Fig. 7-19).

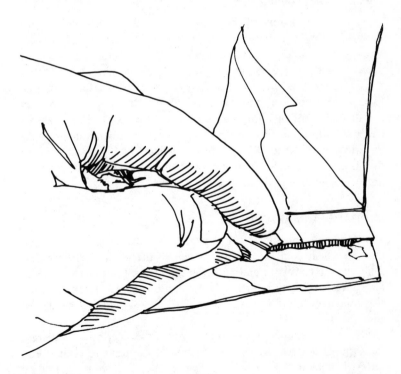

Fig. 7-19. Corrosive flux should be removed after soldering. A damp cloth will often do the trick.

Table 7-3. Suggested Flux Solder Combinations.

METAL	FLUX TYPE*	SOLDER	FLOW TEMPERATURE**
Tin	N/C	50-50	421
Stainless Steel	C	50-50	421
Copper & Alloys	C	60-40	374
Iron	C	60-40	374
Galvanized Metals	C	50-50	421
Steel	C	40-60	455
Electrical Connections	N/C	40-60	455

*N/C Noncorrosive, C Corrosive
**Degrees Fahrenheit

Flux and Solder Combinations

Table 7-3 lists some of the possible flux and solder combinations. Keep in mind that in some cases other solder can be substituted, such as 50-50 solder for 60-40. But the best general results will be achieved with the combinations listed. It should also be noted that fluxes cannot generally be interchanged. A corrosive flux cannot be used when a noncorrosive flux is called for. If a noncorrosive flux is called for and a corrosive flux is used, such as in the case of electrical wiring, damage may be the outcome of the project. On the other hand, if a corrosive flux is called for and a noncorrosive flux is used, complete bonding may not happen because the flux was not strong enough to do the job.

SURFACE PREPARATION

The first rule of successful soldering is cleanliness. Everything that comes in contact with the solder — the metals being joined, soldering iron tool tip and even the solder itself — must be free of dirt, grease, oil, paint, rust or corrosion on the surface. Anything else that may interfere with the solder's adhesion ability should be eliminated.

Of course, the type of project you are soldering will dictate the type of surface preparation. Some of the ways to clean metal surfaces include steel wool, sandpaper, file, disk sander, grinding wheel and wire brush (Fig. 7-20). You should use any practical method to bring the surfaces up to a bright and shiny condition.

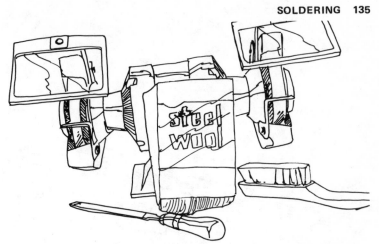

Fig. 7-20. Some common tools and materials used in surface preparation.

Just before you begin soldering, it is also a sound practice to lightly polish the solder itself with either steel wool or sandpaper. This simple act will remove anything on the surface of the solder and insure that you are using a pure solder (Fig. 7-21).

A second rule for successful soldering is that the pieces must fit together well before soldering begins. Don't expect the solder to stand up under a lot of stress and strain because it generally won't. It is far better to have a tight mechanical bond before soldering the work together than it is to expect the solder to do this for you.

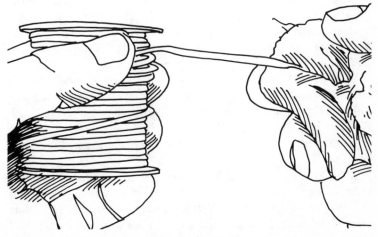

Fig. 7-21. Clean solder with steel wool before using. This will remove surface dirt.

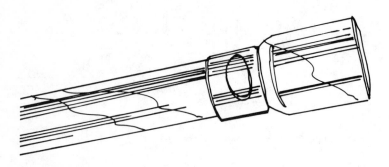

Fig. 7-22. Joints should fit tightly before soldering. A gap of about 0.002 of an inch is acceptable.

In some cases achieving a sound connection or fit will be fairly simple, as in fitting plumbing connections together. But, in other cases, you will have to do quite a bit of bending and hammering to get the pieces to fit together properly before soldering. In all cases, a gap of about 0.002 of an inch is acceptable. If the gap is any larger, the solder may not have the ability to hold the work together. The result may be a failing joint (Fig. 7-22).

Whenever possible you should avoid butt joints, as these are the weakest type in soldering.   Figure 7-23 illustrates some of the more common metal working joints that can be used when soldering. As you can see, most of the joints are of the interlocking type. Fashioning this type of joint requires skill that can only be developed through familiarity with the metals being worked with and time. Nevertheless, there are a few tools that can help you to make some of these joints. These tools include a pair of pliers, a bench vise, hammer and anvil, and your own ingenuity. Remember that the better the metals fit together, the better the finished joint will be (Fig. 7-24).

Keep in mind, when fashioning joints from sheet metal, that this was the way many metals were joined before welding came into existence around the turn of this century. Forming solid joints prior to soldering is good practice for similar work with which will be covered later. In soldering, the better the pieces fit together, the better the finished project can be.

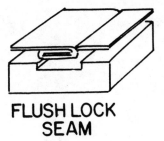

**FLUSH LOCK
SEAM**

**FLANGED
RIVETED**

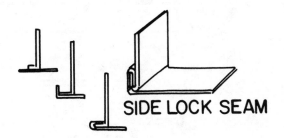

KEYED LOCK

SIDE LOCK SEAM

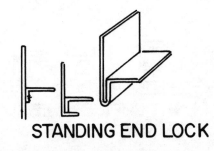

STANDING END LOCK

Fig. 7-23. Soldering joints in common use.

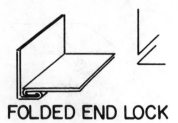

FOLDED END LOCK

INSIDE LOCK

LOCK SEAM

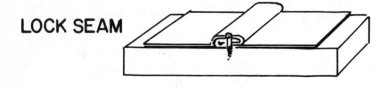

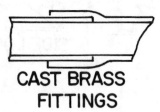

CAST BRASS
FITTINGS

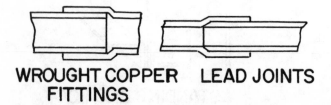

WROUGHT COPPER   LEAD JOINTS
FITTINGS

Fig. 7-23. Soldering joints in common use (cont.).

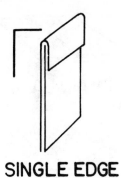
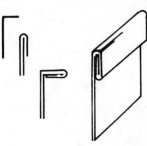

SINGLE EDGE

DOUBLE EDGE

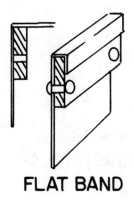
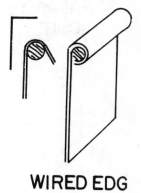

FLAT BAND

WIRED EDG

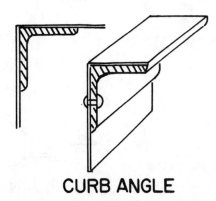

CURB ANGLE

Fig. 7-23. Soldering joints in common use (cont.).

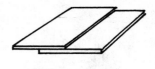

PLAIN LAP

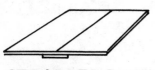

STRAPPED BUTT JOINT

RIVETED LAP

O-GEED

PLAIN LOCK

Fig. 7-23. Soldering joints in common use (cont.).

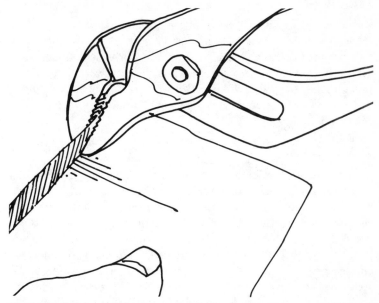

Fig. 7-24. A pair of pliers are very handy for bending sheet metal to form a soldering joint.

Fig. 7-25. Vise-grips are a handy tool to hold places secure for soldering.

TECHNIQUE

Once the proper fit or joint has been established, the piece can be soldered. Before actually beginning, however, it may be necessary to hold the pieces together with a clamp or other suitable tool. This will keep the pieces aligned while you are working on them. It will also be good insurance that the pieces are not disturbed while the solder is cooling, which could result in a weak or ineffective joint (Fig. 7-25).

Tinning the Tip

If you will be soldering with an iron or soldering gun, the first task which must be accomplished before soldering can begin is to heat up the tool and tin the tip. Tinning the tip of a soldering tool simply means coating the tip with molten solder. Since it is impossible to solder with an untinned or badly tinned tip — the oxidized film of copper on the surface of the tip prevents ready transmission of heat — it is important to tin the tip thoroughly. Actually, tinning the tip is a simple process. Begin by heating the soldering tool until it will melt solder (Fig. 7-26). Then simply rub some solder onto the hot tip and spread the molten metal over the surface. When this is properly done, the tip will have a bright, shiny silver appearance.

Fig. 7-26. To tin the tip of a soldering iron, begin by heating the tip and then simply melt solder to cover it.

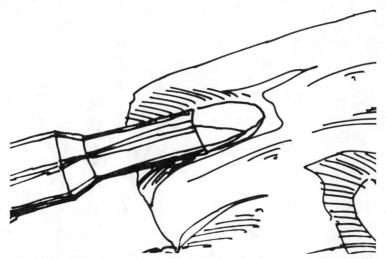

Fig. 7-27. Clean the copper tip of a soldering iron with steel wool before tinning with solder.

If you have a problem getting the solder to stick to the tip, it may be because the tip is dirty. Clean the hot tip with steel wool or a damp cloth; then try again (Fig. 7-27). If the tip is too hot, the copper will tarnish immediately. When this happens, it will be necessary for you to let the tip cool slightly before tinning. You may also find that a little flux rubbed onto the tip before applying the solder will enable you to tin the tip a bit easier (Fig. 7-28). Use either corrosive or noncorrosive, depending on the project.

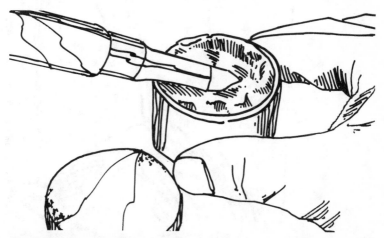

Fig. 7-28. Sometimes it is necessary to apply flux to the soldering iron tip before tinning.

Fig. 7-29. Always heat up the metal with the soldering iron; then apply the solder to the hot joint. Never do it the other way around.

Heating the Work Metal

After the tip of the soldering tool has been sufficiently tinned, you can begin the actual soldering. But first apply some flux to the area being soldered. The proper way to solder is to heat up the work with the soldering tool and then apply the solder to the work. You let the heat of the metal melt the solder rather than melting the solder with the tip of the soldering tool. The obvious reason for this approach is that when the metal is hot enough to melt the solder, you can then expect a reasonably good bond. When the metal is not at the proper temperature, all the solder will do is lie on the surface rather than be bonded to the surface (Fig. 7-29).

There are a number of factors which will determine how quickly you can heat up the work metal to be soldered. These include the thickness of the metal, the size of the tip on the soldering tool — larger tips heat a larger area than small soldering tips — and the power or heating ability of the soldering tool. For small projects which may only require a spot of solder, such as electrical connections, you should use a very small tipped soldering tool. For large soldering projects such as a seam, you will find the work much easier to accomplish if you use a soldering tool with a large tip (Fig. 7-30).

Tinning the Surface

Once the metal is hot enough to melt the solder, your first objective is to tin the surface. The word tinning, in this case, is from an old sheet metal term. It means to apply a very thin layer or film of solder to the metal surface. Your aim is to lay a thin layer along the entire joint or seam, which will cause the pieces to stick together. Then go back over the work and build up the bead of solder until you have a solid, strong and even seam.

In some cases like installing a radiator bracket, both surfaces are tinned before assembly. Then the pieces are pressed together and heated. This causes the mating surfaces, which have already been tinned, to adhere to one another (Fig. 7-31).

As I mentioned earlier, the solder will always flow toward the heat source. You can take advantage of this fact when you need to make solder flow uphill, as when joining copper tubing for an overhead plumbing system or when soldering a vertical seam. Move the heat source slowly along the seam and the molten solder will follow, leaving a solidified trail.

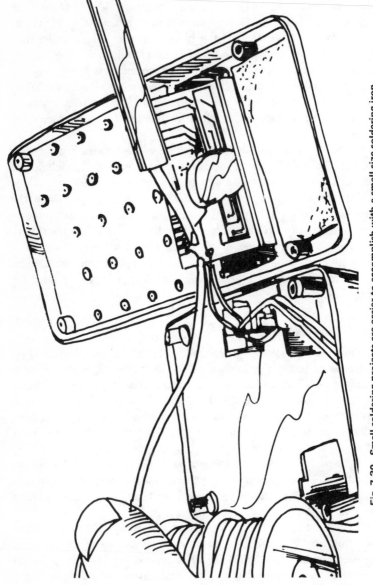

Fig. 7-30. Small soldering projects are easier to accomplish with a small size soldering iron.

Fig. 7-31. Some joints are easier to solder if the mating surfaces are tinned with solder and then fitted together and heated until the solder, on both faces, melts together.

Chunky Solder

When soldering is being done properly, the cooling solder will appear smooth and slightly shiny. If you are moving too quickly, the solder may take on a grayish cast. This means that the metal was not heated quite enough to totally flow the solder. In extreme cases, the solder will appear "chunky," with little sharp edged bits of solder in the bead. This is a problem which indicates that you

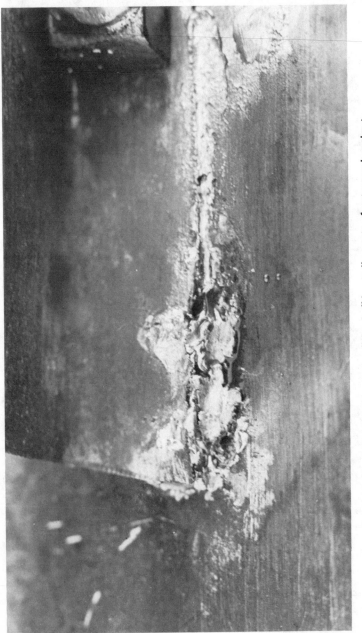

Fig. 7-32. A solder joint that appears "chunky" is usually a case of not enough preheat.

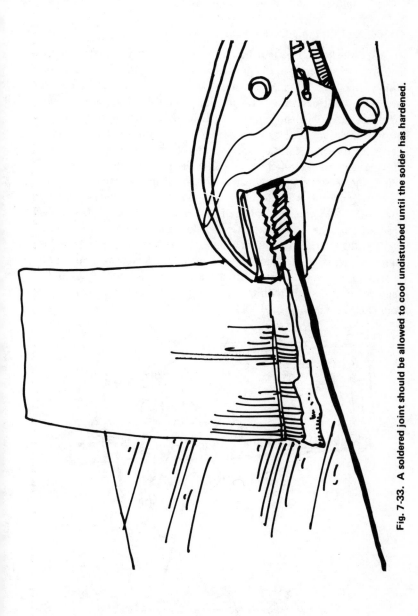

Fig. 7-33. A soldered joint should be allowed to cool undisturbed until the solder has hardened.

are working too quickly and that the solder may be the type that requires more heat to flow properly. For example, 5-95 or 20-80 solder requires higher temperatures than 50-50 solder (Fig. 7-32).

Cooling and Cleaning the Soldered Area

After the joint or seam has been soldered to your satisfaction, remove the soldering tool quickly. Remember that the solder will follow the heat. Once you have removed the heat source, look over the work carefully. If any sharp points are present, or any areas appear not to be covered sufficiently, touch up with the tip of the soldering tool as necessary. When you are satisfied with the joint or seam, let the work sit undisturbed until it has cooled. Moving the still hot metal may cause the seam to come apart, and this is the reason for not touching it until it has cooled (Fig. 7-33).

Most soldering that involves the use of flux will also require that you clean the just soldered area after it has cooled. In the case of noncorrosive flux, there is no danger that the flux will damage the joint or surrounding area. There will be a brown residue that will remain, especially when rosin core solder is used. This brown film is very brittle, actually a pine tar residue, and can be removed quite simply with the tip of a pocket knife or other suitable tool (Fig. 7-34).

Fig. 7-34. You can usually remove noncorrosive flux with the tip of a pocket knife.

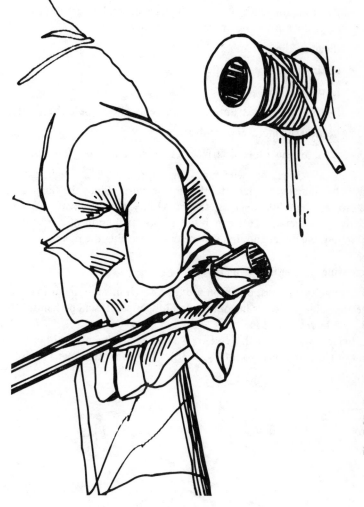

Fig. 7-35. Remove corrosive flux from a cooled soldered joint with a damp cloth.

Corrosive flux must be removed from the finished soldering project for reasons stated earlier. This is usually best accomplished by rinsing the area with water. If this is not possible, use a damp cloth (Fig. 7-35).

Soldering With a Flame

When soldering with a flame, instead of soldering tool with a copper tip, the same basic procedures are followed:

- Cleaning the surfaces.
- Fitting the pieces together.
- Applying flux.
- Heating the metal until it is hot enough to melt the solder.
- Letting the metal cool undisturbed.
- Cleaning the soldered surface of flux and smoothing if necessary.

Generally speaking, working with a torch requires a light touch with the flame and strict attention to the work. Since the flame can be hotter than 800 degrees, the possibility always exists of over-heating the metal. This can result in poor adhesion of the solder, scorching the metal and, in extreme cases, warping the metal.

Soldering Copper Plumbing Pipe

Probably the most common use for a torch in soldering is in completing soldered joints in copper plumbing pipe. It will be helpful to explain how this is done.

Begin by cleaning the end of a section of pipe with either fine emery cloth or steel wool (Fig. 7-36). The same treatment is given

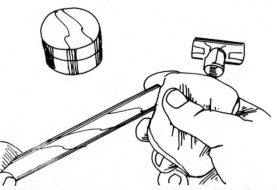

Fig. 7-36. Both the end of copper pipe and the fitting must be clean and shiny. Do this with steel wool.

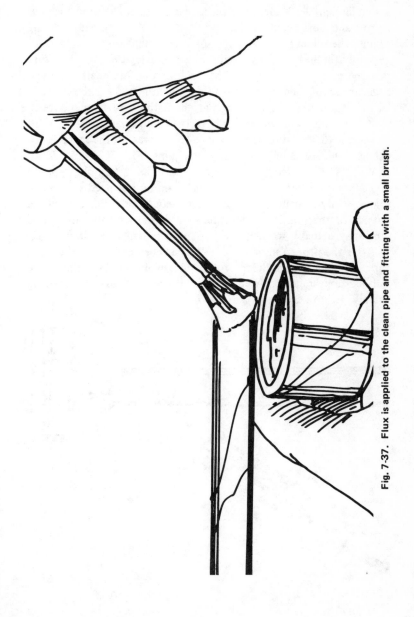

Fig. 7-37. Flux is applied to the clean pipe and fitting with a small brush.

to the fitting. All surfaces should be bright and shiny. Next, apply the flux (Fig. 7-37). In most cases this will be a corrosive paste type, a standard in the trade. The pieces should then be fitted together tightly and positioned so they will not move during the soldering or until the solder has cooled and solidified totally.

Apply the heat to the pipe and fitting, moving the torch to insure that the parts are being evenly heated. As the flux heats up, it will begin to bubble and possibly turn a brownish color. This is usually an indication that the copper is at about the right temperature to melt the solder. Move the flame of the torch away from the joint and touch the end of a strip of solder to the area. If the metal is hot enough, the solder will flow into the joint (Figs. 7-38 and 7-39).

When soldering copper water lines, it is important that you do not apply too much solder. Actually you do not need very much solder. Only a small amount of solder should be visible around the outside of the joint. Excess solder buildup around the fitting and pipe may actually weaken the joint rather than make it stronger.

After the proper amount of solder has been applied to the joint, wipe the area with a damp cloth. This will give the joint a nice finished appearance, as well as clean the area of flux. The wiping will also remove excess solder (Fig. 7-40).

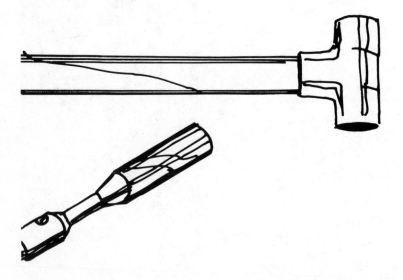

Fig. 7-38. The fluxed fitting is twisted onto the pipe; then heat is applied with a propane torch.

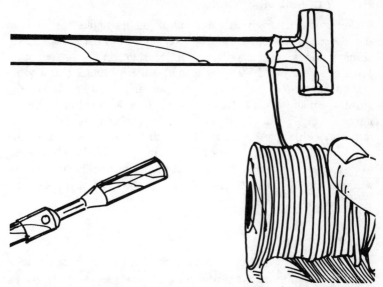

Fig. 7-39. When the pipe and fitting have been heated to the proper temperature, solder is touched to the hot joint and will flow into place, forming a tight seal.

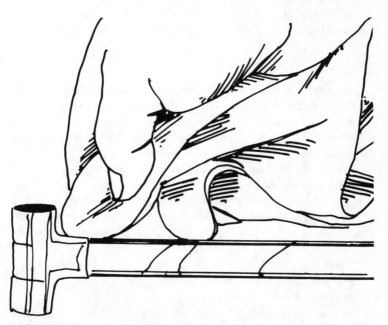

Fig. 7-40. Before the solder has totally hardened, wipe the joint with a damp cloth. Work carefully or you may disturb the joint.

SAFETY CONSIDERATIONS

Soldering is a relatively safe method of joining metals but, as with any operation that involves heat, there is always the possibility of burns, noxious fumes and damage to the skin. In most cases, accidents can be avoided by being careful while working. It is a good idea to have a specific place for the soldering tool when turned on but not actually in use. This can be a wire soldering iron holder as shown in Fig. 7-13, a large ash tray (although these tend to have a rather short life span) or even a brick. If the soldering iron or gun is electrically operated, as most types are, you should set up your work area so there is not any possibility of burning the power cord with the tip of the unit. This could cause a short circuit or even electrical shock.

Eye Protection

When working with corrosive fluxes, it is a good idea to wear a long-sleeved shirt and eye protection. In some cases a pair of gloves may not be such a bad idea.

The chances of mishap increase when you find that you must work away from a workbench. This is often the case when soldering copper plumbing pipe. Precautions should be taken. If working overhead, eye protection and a hat are more than a good idea.

Adequate Ventilation

Whenever you are soldering with a gas-powered torch like propane, you should make certain that there is adequate ventilation in the work area. If you are soldering copper pipes, keep an eye out for areas close to the joint that are combustible such as wooden joists. In many cases, you will be forced to solder very close to these areas. Know that the possibility of fire is always present, and have some means of putting out a fire if one should start. Some good fire protection or fighting tools to have around are a bucket of water, fire extinguisher, a box of baking soda or a plant atomizer. The last item will quickly spray a cloud of water onto a burning joist and extinguish the flame (Fig. 7-41).

If you find it necessary to solder close to wooden joists, slip a piece of asbestos board between the work and the joist. The asbestos will prevent the possibility of a fire.

Since all fluxes give off fumes or smoke when heat is applied, ventilation is more than important. While the noncorrosive rosin

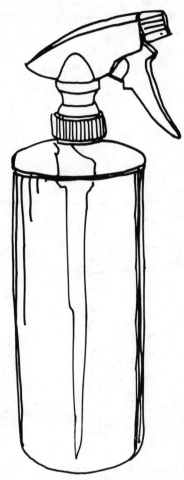

Fig. 7-41. A plant atomizer makes a handy fire extinguisher for overly hot or burning wooden joists. A quick spray will cool the area in a jiffy.

fluxes are not harmful, they may upset your system for a short period of time. It is therefore not a good idea to breath these fumes for long periods. Prolonged inhalation of corrosive flux fumes should be avoided. If the soldering project is a large one, some type of respirator should be worn to protect the worker.

Soldering is an enjoyable means of joining metal which can be accomplished rather inexpensively. It is good practice for the stronger methods of joining metals since some of the techniques are basically the same. Approach all soldering tasks with a bit of caution and develop safe work habits. Not only will this help you to achieve predictable results, but it will help prepare you for other methods of joining metal (Fig. 7-42).

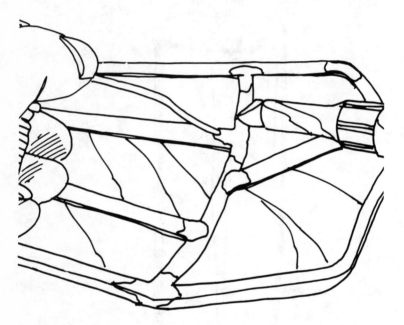

Fig. 7-42. Soldering is an inexpensive and enjoyable means of joining metal.

Chapter 8
Brazing and Braze Welding

In the previous chapter we covered how to join metals by using low heat and a nonferrous filler material. You will recall that soldering takes place at temperatures below 800 degrees Fahrenheit and produces the weakest joint of the three types of metal joining techniques: soldering, brazing and welding. In this chapter we will cover how to join metals, again by using a nonferrous filler material, only in this case at temperatures above 800 degrees. First, it will be helpful to us in our discussion of metal joining to explain and define the terms brazing and braze welding as they are often confused.

USES OF BRAZING

Brazing and braze welding are terms which describe two different techniques for joining metals. Brazing refers to a process where the metal parts are heated to temperatures above 800 degrees Fahrenheit (Fig. 8-1). Then a nonferrous filler metal is introduced into the joint. Part of the key to successful brazing is that the filler metal must have a melting point lower than that of the parent metals. The pieces to be joined are heated up to a point where the brazing filler metal will melt, but not the parent metals.

When brazing metals, the fit between the pieces is very important and must be tight. Since the filler material flows freely over the surface and into the recesses of the joint, it will not fill gaps or other areas which do not fit well. This flowing of the filler material is known as *capillary attraction*. It happens when the contacting surfaces of a liquid (the molten filler material) and a solid (the parent metals) interact so that the liquid surface is no longer flat. A coating of flux aids the capillary attraction and helps the filler adhere to the parent metals.

A brazed joint is stronger than a soldered joint and, in certain instances, is as strong as a welded joint (where the parent metals have been heated up to the melting point and have fused together).

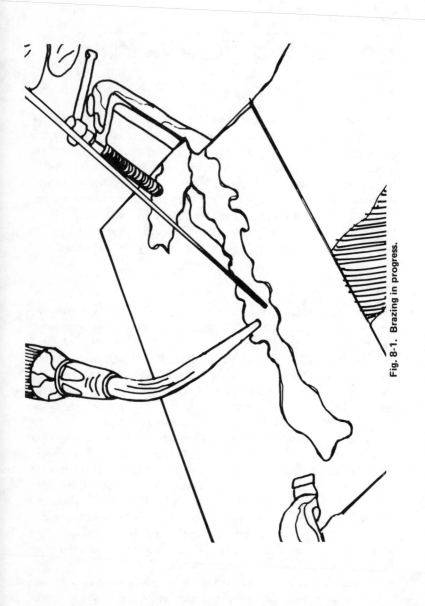

Fig. 8-1. Brazing in progress.

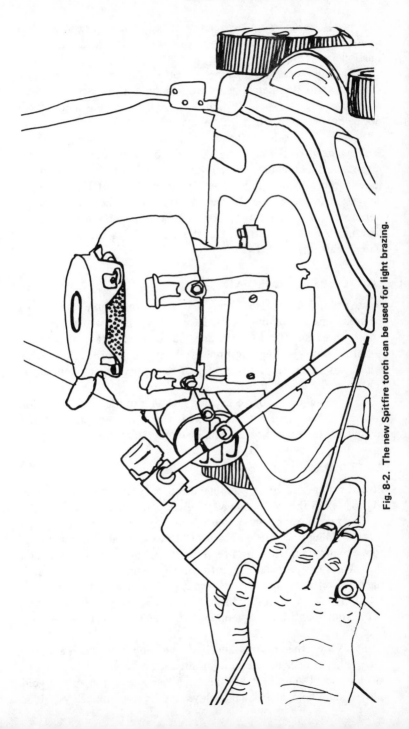

Fig. 8-2. The new Spitfire torch can be used for light brazing.

Brazing is used where mechanical strength and pressure proof joints are required, such as in a pipeline system and cast transmission cases (Fig. 8-2).

Generally speaking, brazing, because of lower temperatures, will not warp the metals being worked on. Brazing is also much quicker and less expensive than welding because the metals need not be heated up to the melting point.

Brazing is a suitable means of joining many types of metals such as steel, cast iron, nickel, copper, brass, bronze, cast bronze, magnesium, aluminum, monel and inconel. For success when joining these metals, it is necessary to carefully fit the pieces together. Use the proper flux and filler material and work at a predetermined temperature. Brazing is also a good technique for joining dissimilar metals such as the nickel alloys and copper alloy base metals.

USES OF BRAZE WELDING

Braze welding differs from brazing in the design of the joints being secured. Braze welding is used for groove, fillet and other weld joints where the filler material does not take advantage of capillary action. Braze welding does, however, use a nonferrous filler metal which has a melting point below that of the parent metal but above 800 degrees Fahrenheit.

The technique for braze welding is similar to that used for fusion welding. The important exception is that the parent metal is not melted but only raised to the tinning temperature.

FILLER MATERIAL

Both brazing and braze welding use a nonferrous filler material. In most cases this will be an alloy of copper, zinc and copper or tin.

The two most commonly used copper alloys for brazing and braze welding are brass and bronze. Brass is an alloy consisting of copper and zinc, while bronze is comprised mainly of copper and tin. The more popular for both joining techniques is brass. It is slightly interesting to note that even experienced welders call brazing and braze welding rods "bronze" when actually they are using brass rods consisting of copper, zinc and about 1 percent tin (Fig. 8-3).

There are other brazing and braze welding rods which are used for joining special metals in addition to the copper alloys. These include nickel and chromium alloys, silver alloys, copper and gold alloys, aluminum and silicon alloys and magnesium alloys. It is

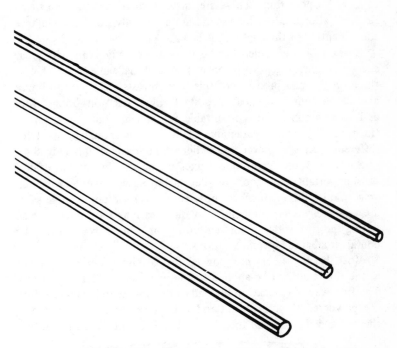

Fig. 8-3. Three common sizes of "bronze" rod.

probably a safe assumption that the do-it-yourself welder will use bronze rod for most types of brazing and braze welding. This is a good nonferrous filler material for joining cast iron and steel. For joining other types of metals, it is always a good idea to consult your local welding supply house for the best type of nonferrous filler material.

FLUX

Flux plays a very important role in braze welding and brazing. Most brazing processes require the use of a flux to prevent oxidation during the application of heat. A flux, you may remember, is used in soldering to protect the metal surfaces being joined from the atmosphere. In brazing and braze welding the flux is used for the same reason. In addition, a flux will aid the capillary attraction of the nonferrous metal while brazing at the same time, dissolving the oxides that may form during either brazing or braze welding.

Flux is commonly available in powder, paste and liquid forms. Also available are bronze rods that have been coated with a flux which eliminates the need for additional flux. There is no such thing as an all-purpose flux. There are, instead, fluxes designed for specific applications. Most fluxes sold are chemically pure and will have some mention of this fact on the label.

Borax is a flux which has been used for many years with great success for brazing brass, bronze, steel and malleable iron. One of the more popular mixtures of this flux consists of 75 percent borax and 25 percent boric acid in paste form. This "anti-borax" flux as it is sometimes called is suitable for brazing those metals mentioned earlier in this paragraph. Other fluxes should be used for different metals, however, since there is no universal flux (Fig. 8-4).

Since the fumes from most fluxes can be harmful to your health, it is important that good ventilation be present in the workshop area. Some fluxes, such as those containing sodium cyanide salts, are very dangerous. You should avoid breathing the fumes or even letting the flux come in contact with your skin. Always check the container label for specific cautions.

Most fluxes are sold in a powder form. This is quite convenient as the flux can then be applied to the metal in two different ways. The first way is to heat up the end of the bronze rod and dip it into the powdered flux. The hot rod will cause the flux to adhere to the rod and the rod can then be used as it is, with a coating of flux

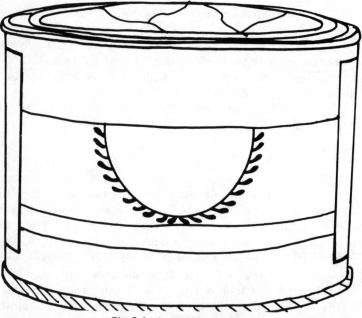

Fig. 8-4. A good brazing flux.

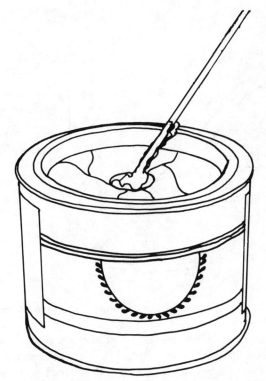

Fig. 8-5. A hot brazing rod is dipped into the flux and is quickly coated.

(Fig. 8-5). Powdered fluxes can also be made into paste type fluxes, usually by adding water until the right consistency is achieved (Fig. 8-6). Paste fluxes are commonly applied to the metal with a small brush. It is important that you read the container label as some fluxes require that you mix with alcohol instead of water to form a paste.

PREPARING SURFACES

Easily, 75 percent of the success of any brazing project is directly related to the condition of the surfaces being joined. Not only must the surfaces be physically and chemically clean, but the pieces must fit together precisely. A joint clearance of 0.002 to 0.006 of an inch is acceptable when copper alloy or silver alloy brazing. When joining aluminum, greater tolerances are permissible, 0.006 to 0.015 of an inch. As you can imagine, joint design is quite important and will be discussed in this chapter. The clearances of a joint will have a direct bearing on the strength of the finished project. There must be a

Fig. 8-6. Mix up water and powdered flux to form a paste.

certain amount of clearance so the nonferrous metal filler can be drawn by capillary action into the joint. If the clearance is too great, however, the capillary attraction will be lessened. The end result will be a joint that is weak.

Cleaning Metal

There are several ways of cleaning metal before the pieces are fitted together for brazing. These include wire brushing, emery cloth, crocus paper, steel wool and, in special circumstances, washing the parts with an acid solution. If the metal has just been machined or milled, all that is usually required is a light touch with steel wool. But the older the metal, in relation to elements, the more surface preparation that will be required.

If rust is in evidence, for example, you should begin cleaning with a wire brush (Fig. 8-7). In most cases this tool will remove all of the surface scale and oxidized metal. Next, the metal should be sanded with emery cloth or crocus paper to make the surface smooth and shiny.

It is generally not recommended that a grinding wheel or other harsh abrasive be used on metal that will be brazed. The main reason is that these tools have a tendency to impart scratches or a rough texture to the surface. Knowing that close tolerances between

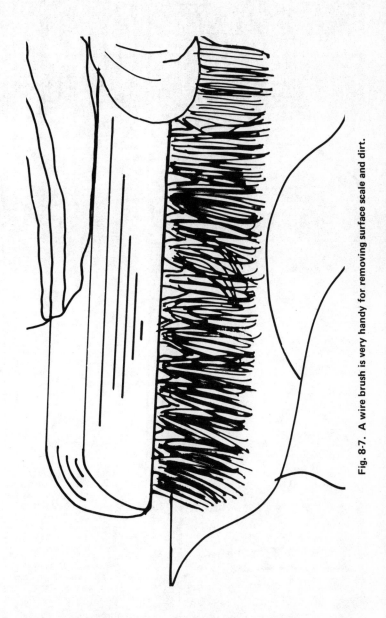

Fig. 8-7. A wire brush is very handy for removing surface scale and dirt.

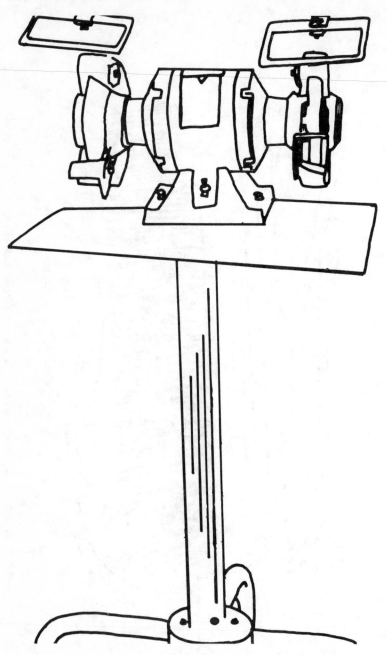

Fig. 8-8. While a grinding wheel is not generally suitable for metal that will be brazed, it can be quite useful for metal that will be braze welded.

the metal parts are important, this type of cleaning may cause tolerances which are unacceptable, at least from the standpoint of brazing.   A grinding wheel is suitable for surface preparation of metals that will be braze welded (Fig. 8-8).

If much scale is in evidence on steel or iron and it cannot be easily removed by wire brushing, steel wool or emery cloth, it may be necessary to soak the pieces in a special *pickling bath.*   This is simply a mixture of water and sulfuric acid.   After the pieces have soaked in the bath, they must be rinsed in warm water until all of the acid residue is flushed away.   Since working with acid can be dangerous, you should proceed with due caution.   Your local welding supply house should have the stock ingredients for a pickling bath and can probably offer some suggestions as to how you might best go about this method of cleaning steel and iron.   They may even perform this service for you.   Generally speaking, a pickling bath will clean the metal like no other method.   It is a useful means of cleaning both small pieces as well as those metals which might require an enormous amount of work to clean.

Butt and Lap Joints

After the metal has been cleaned, the pieces must be fitted together so the tolerances are close, 0.002 to 0.006.   The best way to make certain that the pieces fit together this closely is to use or develop a joint that is tight fitting.   The two most common joints used are the *lap joint* and the *butt joint.*   Whenever possible, the former is much preferred.

The main reason that the lap joint is popular when brazing metals is that it offers the strongest joint.   It is also relatively easy to make.   Simply lap two surfaces together and braze.   The length of the lap should be equal to at least three times the thickness of the thinnest member being joined.   For example, if you were brazing two pieces of 3/8-inch thick steel, using the lap joint, the length of the lap should be at least 1 1/8 inches.   Figure 8-9 illustrates some of the more common joints, both good and bad.

Applying Flux

After the pieces have been cleaned thoroughly, the joint surfaces must be coated with the proper flux.   Remember that there are different fluxes.   You must choose the right one for the metal you are working with.   In most cases a paste type flux is brushed on the surfaces.   If you are using a powder flux, follow the directions on

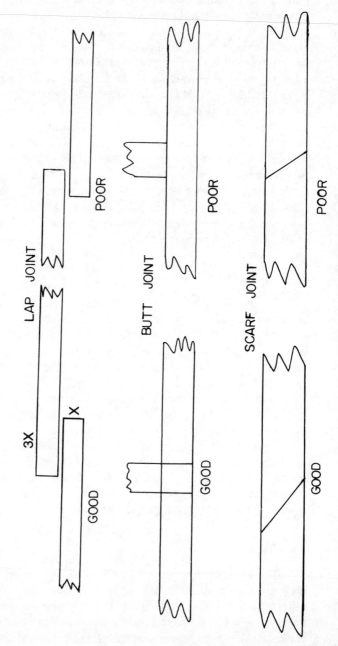

Fig. 8-9. The three most common joints used in brazing.

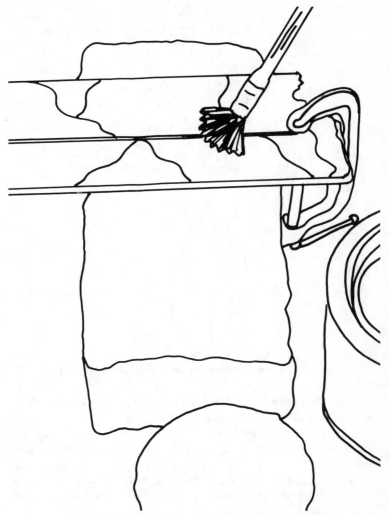

Fig. 8-10. Paste flux is brushed on a joint to be brazed.

the container to form a paste and brush this on the joint surfaces (Fig. 8-10).

Once the pieces have been fluxed and fitted closely together, they must be held in this position until they have been brazed and the filler metal has cooled to the solid stage. In many cases a C-clamp can be used for this. As an alternative, a pair of locking jaw pliers are quite handy. There are many different sizes and shaped ends available. Your local welding supply house will probably have a selection of these, as they are quite popular with professional

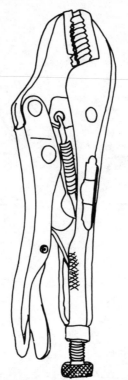

Fig. 8-11. Vise-grips have many uses around the welding shop.

welders. A standard 6 to 8-inch pair of locking pliers has hundreds of uses around the welding shop (Fig. 8-11).

The next step in the brazing process is, of course, joining the pieces with heat and filling the space between with molten non-ferrous metal. Earlier in this chapter I mentioned that careful surface preparation was about 75 percent of the task of brazing. The remaining 25 percent involves technique, which is what we will deal with now.

BRAZING TECHNIQUE

Since brazing is decidedly an oxyacetylene process, the directions for this method of joining metal will be discussed in oxyacetylene terms. As with all oxyacetylene metal joining, it is important when brazing to choose the correct tip size for the project at hand. The type and thickness of metal being worked on will determine the correct tip size to use. Since some metals like aluminum can be heated quite easily, a small tip should be used. It is important to keep in mind that the intention in brazing is not to melt the parent

Table 8-1. Welding Tip Sizes and Pressure Settings for Brazing.

METAL THICKNESS	TIP SIZE	SIZE OF BRAZING ROD	OXYGEN PRESSURE (psi)	ACETYLENE PRESSURE (psi)
1/32	1	1/16"	5	5
3/64	2	1/16	5	5
1/16	3	1/16	5	5
3/32	4	3/32	5	5
1/8	5	3/32	5	5
3/16	6	3/32	6	6
1/4	7	1/8	7	7
5/16	8	5/32	8	8

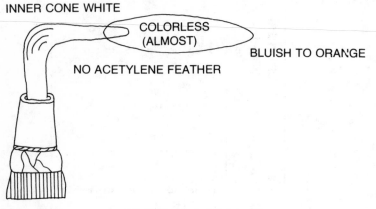

INNER CONE WHITE

COLORLESS
(ALMOST)

BLUISH TO ORANGE

NO ACETYLENE FEATHER

Fig. 8-12. Neutral flame.

metal, but to simply get the joint edges hot enough to melt the filler material and make it flow. You should also keep in mind that a neutral flame has about the same temperature with all size tips, but you may be able to have more control with a small tip. Probably the best place to look for the proper size tip for brazing is in the handbook that came with your torch. Here the manufacturer will have offered suggestions for tip sizes based on type and thicknesses of the various metals. See Table 8-1.

Light the torch and develop a neutral flame which is about best for brazing. Some authorities suggest that a slightly oxidizing flame is better for brazing. In time you may feel the same. I suggest when you are relatively new to working with a torch that you work with a specific flame until you have mastered the basics of flame and torch control (Fig. 8-12).

Heating the Base Metal

The whole key to successful brazing is heating the base metal up to a temperature that will be hot enough to melt the nonferrous filler metal, but not cause the parent metal to become liquid. In most cases this is not as difficult as it may sound, as the melting point of all copper alloy brazing rods, brass and bronze, is between 1,100 to 1,800 degrees Fahrenheit. The melting point of most steels and iron is at least 2,100 degree Fahrenheit.

It is important that the joint be evenly heated. To accomplish this, it is almost always best to keep the torch moving in a circular, half-moon or zigzag motion over the area being heated (Fig. 8-13).

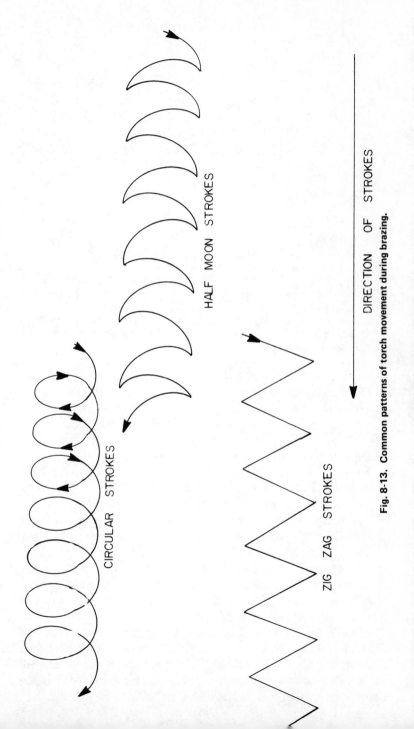

Fig. 8-13. Common patterns of torch movement during brazing.

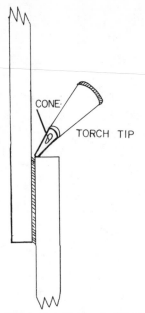

Fig. 8-14. It is important to keep the cone of the neutral flame close but not touching the parent metal. Keep the torch moving over the parts of the joint to distribute the heat evenly.

If the joint is a long one, heat up a section of it at one time. Then proceed along the joint after the initial section has been brazed.

As you heat the metal, watch the flame closely through welding goggles, of course. You will want to keep the cone of the neutral flame close to the edges of the joint but not touching the metal. Since the tip of the cone is the hottest part of the flame, it may melt the metal if it comes too close to the joint. Overheating is a common problem when learning how to braze (Fig. 8-14).

As the metal heats up, the flux around the joint will give you good indications of the approximate temperature. This is another good reason for choosing the proper flux for the metals being brazed. When heated properly, the flux will boil at about 212 degrees Fahrenheit, leaving a whitish powder on the metal surface. This powder will become puffy and begin to bubble at around 600 degrees Fahrenheit. It will become a liquid on the surface at about 800 degrees Fahrenheit. At a temperature of about 1,100 degrees Fahrenheit the flux will once again become liquid and darkish in color. This indicates that just a bit more heat is required to melt the filler rod. The metal of the joint, at this point, will have a dull red color which is another indication that the temperature is about right to melt and flow the filler rod.

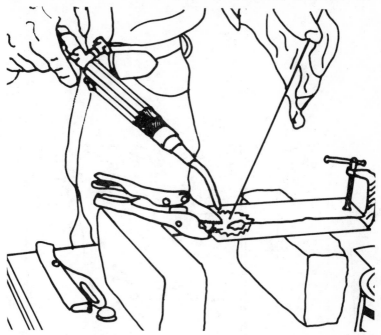

Fig. 8-15. Apply the brazing rod only after the metal surfaces have been heated up to the point where they will melt the rod.

Bonding

When the metal is at approximately the right temperature, move the flame away from the joint and touch the end of the filler rod into the joint. If the metal is hot enough, the rod will flow over the heated area and into the joint (Fig. 8-15).

It is important to keep in mind that both pieces of metal must be of equal temperature or the filler rod will flow only to the hottest piece. This is the main reason for moving the flame over the joint in a circular (or other suitable) motion.

Once the filler metal flows over the heated area, bonding will take place. You should not disturb the braze until the metal has cooled. This will insure that the bond is a sound one.

After the joint has cooled, you may want to clean up the braze. The amount of cleaning that will be necessary will depend on the parent metals, the filler nonferrous metal and the flux used. In most cases, only a light slag or discoloration will be present on the brazed joint. The discoloration can be easily removed with special cleaners or a wire brush. If the slag buildup is heavy, you may have to resort to a chipping hammer and possibly a cold chisel to clean the slag off the joint (Figs. 8-16 through 8-18).

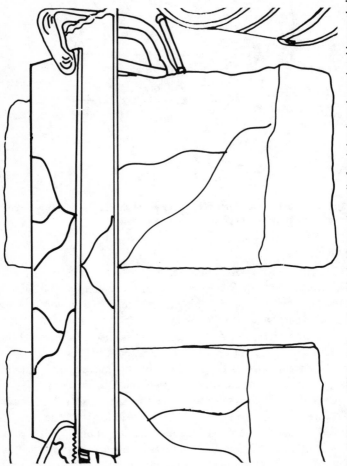

Fig. 8-16.   The work to be brazed must be held securely until after it has been brazed and has cooled.

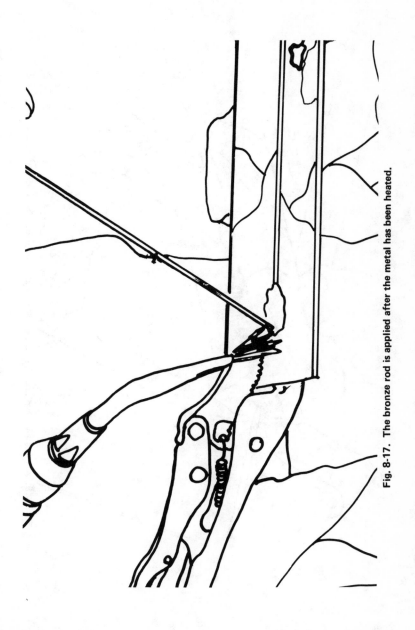

Fig. 8-17. The bronze rod is applied after the metal has been heated.

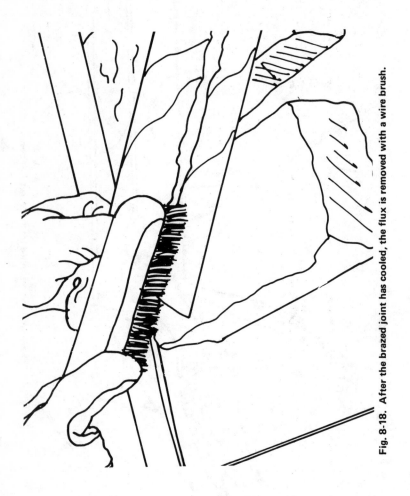

Fig. 8-18. After the brazed joint has cooled, the flux is removed with a wire brush.

BRAZING VERSUS FUSION WELDING

The obvious main advantage of brazing over fusion welding is that it is the best way to join different or dissimilar metals. Just about any two metals can be joined by brazing except aluminum and magnesium. Another strong point in favor of brazing is that since the metals need not be brought up to the melting temperature of the parent metals, there is almost never any distortion of the metal. This holds true when brazing is correctly done on any thickness of metal.

SILVER AND BRONZE BRAZING

In most instances you will be brazing with a bronze filler rod. One alternative to this nonferrous metal is silver rod. Silver brazing differs only slightly from bronze brazing in that the melting temperature of the silver is lower than bronze. This will mean that less heat is required for the joining and the brazing can be accomplished a bit quicker. Silver rod, as you might expect, is more expensive than bronze. The technique for silver brazing is essentially the same as for bronze brazing. A quick outline of these procedures may be handy for you at this time.

- Thoroughly clean the joint surfaces.
- Apply suitable flux to the joint area and only where you want the silver to flow. Silver rod will not adhere to any surface that is not flux coated.
- Fit the pieces together and hold them in place.
- Heat the joint area evenly.
- Apply the appropriate brazing alloy (in this case silver rod).
- Let the braze cool undisturbed.
- Clean the surface.

BRAZING ALUMINUM

Aluminum can also be brazed in the same manner. Special aluminum brazing filler material and flux must be used for this, however, as brazing aluminum takes place at about the lowest temperatures of all brazing, 815 to 1,220 degrees Fahrenheit. Aluminum brazing requires the same general surface preparation procedures as when cleaning other types of metal. Either mechanical or chemical cleaning should be completed before attempting to braze aluminum. Many experts recommend chemical cleaning for aluminum because it is more thorough. Chemical cleaning consists of a series of steps which include dipping the pieces into a caustic bath, rinsing, dipping once again into an acid bath and then rinsing. This chemical cleaning is repeated until the surface is totally clean. As with the acid pickling solution mentioned earlier, the ingredients for chemically cleaning aluminum should be available from your local welding supply house.

Since brazing aluminum is done at relatively low temperatures, a small tip size is often the best to use. This will give you more control, but not necessarily less heat. It is therefore very important to keep the flame constantly moving over the area being heated.

BRAZE WELDING TECHNIQUE

The two major differences between brazing and braze welding are the joints between the pieces and the fact that braze welding does not utilize capillary action. Joint design for braze welding is similar to those joints used in fusion welding. Although the pieces must fit together snugly, the close tolerances necessary for successful brazing are not necessary when joining metals with the braze welding process. In effect, the filler metal, which is a nonferrous material as in brazing, is used to fill in the gaps between the pieces. When the filler metal comes in contact with the hot parent metal, it tries to open up the grain. As the filler seeps into the grain of the parent metal, a strong bond is achieved. To be sure, the pieces must fit together well, but not necessarily to the close tolerances required in brazing.

Beveling Metal Edges

Braze welding is a popular means of joining many types of metal using the butt joint. When joining metal that is thicker than 1/4 inch, it is necessary to bevel the edges of the joint (Fig. 8-19). For metal of lesser thicknesses, it is not necessary to bevel the edges. They should be chipped or ground slightly until bright metal shows. This will aid the filler in adhering well.

Probably the easiest and quickest means of beveling metal edges is on a grinding wheel (Fig. 8-20). You must work carefully, however, as grinding wheels remove metal quickly in most cases. It will be helpful if you mark the pieces and the angle of bevel with a crayon or other suitable marking tool first (Fig. 8-21).

Flame Cleaning

All parts should be clean of any paint, grease, oil or any other foreign matter. Solvents and wire brushing will usually do the job.

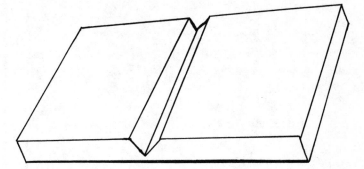

Fig. 8-19. Metal 1/4 inch thick and thicker must have beveled joint edges when braze welding.

Another method used to clean metal of surface dirt is called *flame cleaning* (Fig. 8-22). This is simply a process where a neutral flame is passed quickly over the surface of the metal and burns off the dirt. This is not a recommended cleaning procedure for metal which has a coating of oil or grease, however, as these will burn violently in the presence of pure oxygen.

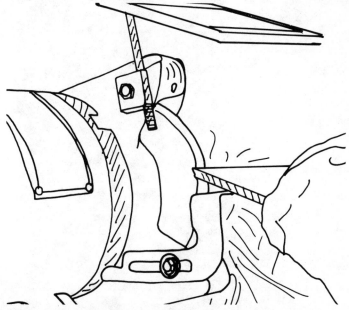

Fig. 8-20. Use a bench grinder to bevel the edges of metal which will be braze welded.

Fig. 8-21. Mark metal with a soapstone pencil before grinding. Use the mark as a guide to the proper bevel angle.

Fig. 8-22. Flame cleaning metal.

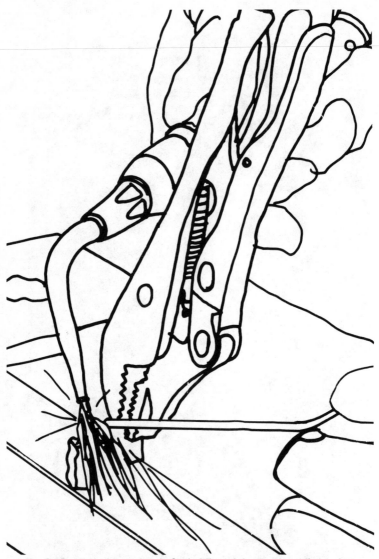

Fig. 8-23. Vise-grips are handy for holding a joint steady while brazing.

After the pieces to be joined have been cleaned, they must be aligned and held in this position until after braze welding. This can be accomplished in the same manner as when holding pieces in a position prior to brazing, with clamps or locking pliers (Fig. 8-23). On large projects, the pieces are often tack welded at this point (Fig. 8-24).

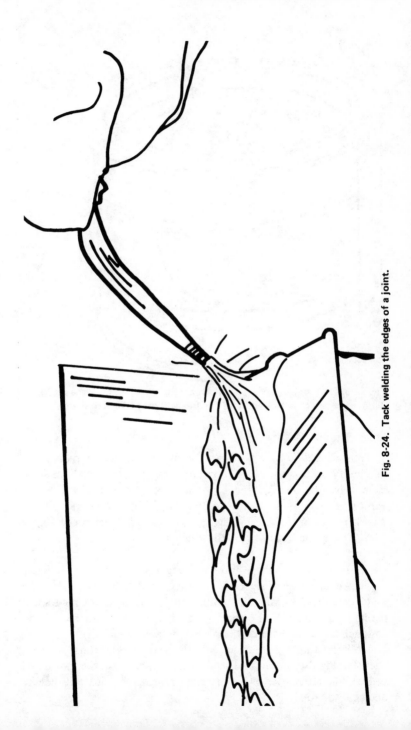

Fig. 8-24. Tack welding the edges of a joint.

Fig. 8-25. To coat a brazing rod with flux, simply heat the tip of the rod. Then dip it into a can of powdered flux. The flux will then adhere to the rod.

Preheating and Heating

Preheating is often necessary in braze welding to reduce the possibility of expansion and contraction stresses which may be caused by the heat of the braze welding operation. Preheating can be accomplished with a neutral flame. When the metal turns a dark red color, it is about right. Keep in mind that preheating is usually only necessary with the thicker pieces of metal, say over 1/4 inch.

It is important when braze welding to keep the torch flame moving over the surface of the joints. This will insure that the metal is heated evenly along the joint. The flame as mentioned earlier is neutral, and the cone of this flame should not be allowed to touch the metal because the parent metal may melt.

While heating the joint surface, the end of the bronze welding rod should also be heated and then dipped into the powdered flux. When the bronze rod is hot enough, the flux will adhere and coat the rod (Fig. 8-25).

Fig. 8-26. A thin coat of bronze on a metal surface.

Tinning

When the parent metal is a very dark red, it is about time to push the end of the bronze rod into the joint. This will allow some of the flux to melt onto the surfaces. As the parent metal melts the bronze rod, it will form a thin coat on the metal surfaces (Fig. 8-26). This is known as *tinning* and is similar to tinning in soldering.

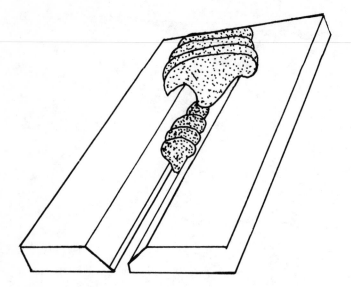

Fig. 8-27. Joining thicker metal with two-pass brazing.

When tinning the surfaces of the joint in braze welding, you will find a rubbing action the best way to get the flux and bronze to stick to the metal.

You must be careful not to heat the parent metal too much as this will cause the filler metal to bubble. Good tinning will then be impossible. On the other hand, if the metal is not heated enough, the filler metal will form little balls on the surface and not in the surface.

Once the surfaces of the joint have been completely tinned, you can go back over the joint and add more filler metal in one or more passes. It is very important that you work continuously, tinning and then building up the filler metal. If the filler metal is allowed to cool, it may break down when reheated. The end result will be a very poor quality braze weld (Fig. 8-27).

Cooling

Even cooling is important when braze welding. There are several ways of accomplishing this, including heating the parent metal around the joint at least twice as wide as the distance of the joint. For example, if the width of the joint is 1/2 inch, heat the parent metal on both sides of the joint for a distance of about 1 to 1 1/2 inches away from the joint. Metal heated in this way will cool slower than if just the joint area is heated.

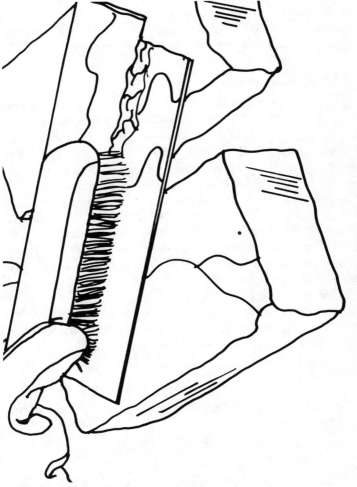

Fig. 8-28. Wire brushing will remove flux after brazing.

Another way to slow down and make the cooling more uniform is to cover the welded piece with asbestos cloth. This will cause the piece to cool very slowly.

Still another way of cooling, and a method often used in industry when working with cast iron, is to pack the hot project in ashes or lime. This will almost guarantee that the metal cools very slowly, thus creating more ductility in the welded material.

After the metal has cooled, it may be necessary to clean up the welded area. This is done to remove any flux residue or slag. A wire brushing is usually all that is required for this cleaning (Fig. 8-28).

GENERAL COMMENTS

Brazing and braze welding techniques are relatively easy to learn and master, providing the do-it-yourself welder devotes the necessary time to surface preparation and flame control. Undoubtedly the best way to learn how to braze and braze weld is to practice on scrap pieces of steel until you have a clear picture of what is necessary for success. Since brazing is, in effect, an extension of soft soldering, this may be a good place to learn the basics. Begin by learning how to make simple lap joints using a soldering iron, flux and solder. Work on sheet metal until you have an insight into the subtleties of heating metal. Once you have a handle on this method of metal joining, move on to thicker metal and use an oxyacetylene torch to braze the pieces together.

In the beginning, it may be easier to work on the thinner gauges of steel, say 1/8-inch thickness or less. Here again you should practice the lap joint as it is probably the easiest to master. Keep in mind the basics of successful brazing.

— Clean the surfaces to be joined.

— Fit the pieces together closely and use the proper flux.

— Heat the metal evenly by constantly moving the neutral flame over the joint area.

— Keep an eye on the flux as this will be one of your best guides as to the temperature of the metal.

— When the metal is hot enough, remove the flame and introduce the brazing rod. It should flow into the joint because of the temperature of the metal and not from the heat of the flame.

— Let the brazed joint rest undisturbed until cool.

— Clean the cooled joint with a wire brush to remove flux and slag, if necessary.

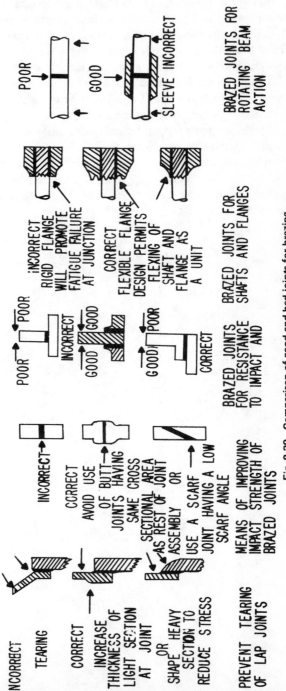

Fig. 8-29. Comparison of good and bad joints for brazing.

After you have mastered the light touch necessary for successful brazing and the lap, butt and scarf joints, move on to braze welding. You must remember that part of the key to braze welding is applying a thin coat of the nonferrous filler material to the surfaces of the joint, known as tinning the surface. It is very important that this tinning coat be soundly attached to the surfaces of the joint metal. This requires that the surfaces be brought up to the proper temperature before applying the filler rod.

In the beginning your brazed joints may not stand up under strain. As you become more familiar with the various metals and temperature ranges, you should be able to produce professional looking finished joints (Fig. 8-29). Once you have mastered the techniques of brazing and braze welding, the obvious next step is to learn how to weld.

Chapter 9

Beginning Welding

Up until this point, we have discussed the two methods used for joining metal: soldering, which takes place at temperatures less than 800 degrees Fahrenheit; and brazing, which requires flame temperatures from 800 to just below the melting point of the metal being joined. In both cases, the metals being joined were not brought up to a temperature which exceeded the melting point of that particular metal. Now we will deal with the final method of joining metals and the one which produces the strongest attachment: welding.

FUSION WELDING

For our purposes, the term welding is rather loosely used to describe the process whereby two pieces of similar metal are heated up to their particular melting point. The molten metal then flows together forming one homogeneous piece. In actuality, this process is called *fusion welding*.

A sound welded joint is a thorough mixture of the base metal and, in many cases, a filler rod made from a similar material. Regardless of the thickness of the metal, a good welded joint will be fused throughout the joint. When joining relatively thin pieces of metal, less than 1/4 inch thick, a thoroughly welded joint is not difficult. But when joining thicker metals, special joint preparation techniques must be employed. As with both soldering and brazing, joint preparation is a necessary part of fusion welding.

Obviously, the best way to learn how to weld is to practice often and on various thicknesses and types of metals. It must be assumed, at this point, that you are familiar with your oxyacetylene equipment. You should know, for example, how to set up the equipment and develop the various types of oxyacetylene flames. These are the acetylene flame, carburizing flame, neutral flame and oxidizing flame (Fig. 9-1). You should also have developed your

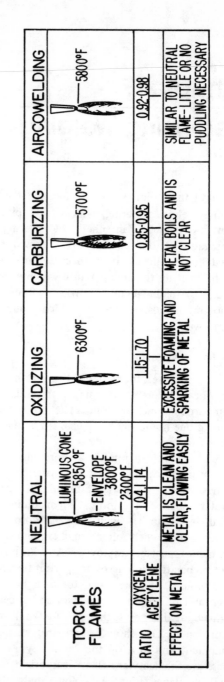

TORCH FLAMES	NEUTRAL	OXIDIZING	CARBURIZING	AIRCOWELDING
	LUMINOUS CONE 5850 °F ENVELOPE 3800°F 2300°F	6300°F	5700°F	5800°F
RATIO OXYGEN ACETYLENE	1.04-1.14	1.15-1.70	0.85-0.95	0.92-0.98
EFFECT ON METAL	METAL IS CLEAN AND CLEAR, FLOWING EASILY	EXCESSIVE FOAMING AND SPARKING OF METAL	METAL BOILS AND IS NOT CLEAR	SIMILAR TO NEUTRAL FLAME—LITTLE OR NO PUDDLING NECESSARY

Fig. 9-1. Characteristics of oxyacetylene welding flames.

own technique for keeping the torch flame in motion and at a specific height above the work. These techniques were discussed in Chapter 8, but a quick review may be of help at this point.

TORCH MOVEMENT

In order to insure that the metal or joint between the two pieces of metal is heated evenly, the flame of the torch must be kept moving so that the metal will be brought up to a uniform temperature and will flow at the same time. This is best accomplished by moving the torch tip continuously. Figure 9-2 shows several of the methods or patterns, if you will, of moving the torch to distribute the heat evenly. At this time, at least one of these torch movement techniques should be familiar to you. More often than not, I find a circular movement the most effective and easiest to perform.

When welding metal, there is another technique which you will have to perfect. This technique also concerns torch movement, either forward or backwards, along the joint. Actually there are two possibilities for general torch movement: the forehand welding and the backhand welding methods.

When using the forehand welding method, the torch flame moves along the joint towards the tip of the welding rod. The position of the welding rod is actually on the cooler and not yet molten area of the joint. The flame of the torch is pointed in the direction of the welding, but also slightly downward so the flame preheats the joint area (Fig. 9-3). The forehand method works well with thin, sheet type metal.

The backhand welding method is opposite from the forehand welding method. Actually the flame of the torch precedes the welding rod in the direction the weld is being made. When back-

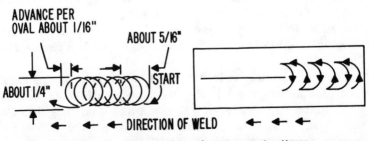

Fig. 9-2. Two common blowpipe movement patterns.

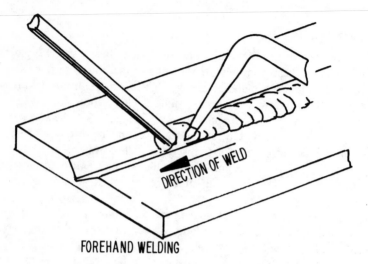

FOREHAND WELDING

Fig. 9-3. Forehand welding technique.

hand welding, the welding rod is used to work the puddle of molten metal. As you can see, from (Fig. 9-4), this differs significantly from the forehand method.

Generally speaking, the backhand method of welding offers greater control of the welding process while at the same time providing welds with equal strength. In most cases it is the preferred technique for welding metal thicker than 1/4 inch.

Throughout this chapter, unless otherwise stated, the flame being used for the welding process is a neutral flame. You may recall that the new neutral flame is the third type which is normally developed when a torch is fired up (Fig. 9-5).

As I mentioned earlier, the best way to learn how to weld well is to weld often. In the beginning of your welding practice work on thin pieces of metal, 1/8 inch or so thick. After you have mastered the various techniques necessary to weld metals of this thickness, move up to thicker pieces. Welding thicker metals is discussed in Chapter 10.

One very good way to learn some of the basics of heating metal up to the melting point is to practice on a single piece of sheet steel. In the beginning, work with steel which is no more than 1/8 inch thick. This metal is easily heated and is quite suitable for learning some of the techniques of welding.

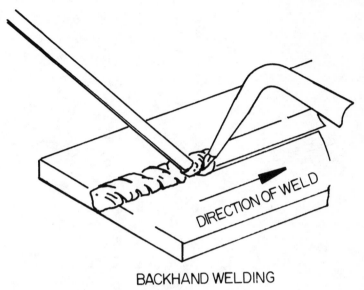

DIRECTION OF WELD

BACKHAND WELDING

Fig. 9-4. Backhand welding technique.

PUDDLING METAL

As your first exercise, begin by placing a 6 to 8-inch long piece of steel between two firebricks (Fig. 9-6). Set up your welding equipment and attach a small welding tip to the blowtorch handle. Consult the guide which came with your unit for the proper oxygen and acetylene settings for the tip being used. Light the torch and develop a neutral flame. It is assumed, at this point, that you have put on suitable clothing, gloves and tinted goggles.

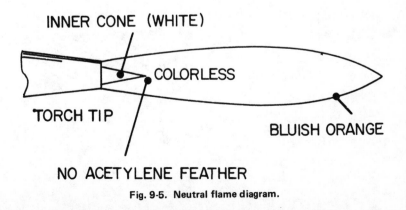

INNER CONE (WHITE)

COLORLESS

TORCH TIP

BLUISH ORANGE

NO ACETYLENE FEATHER

Fig. 9-5. Neutral flame diagram.

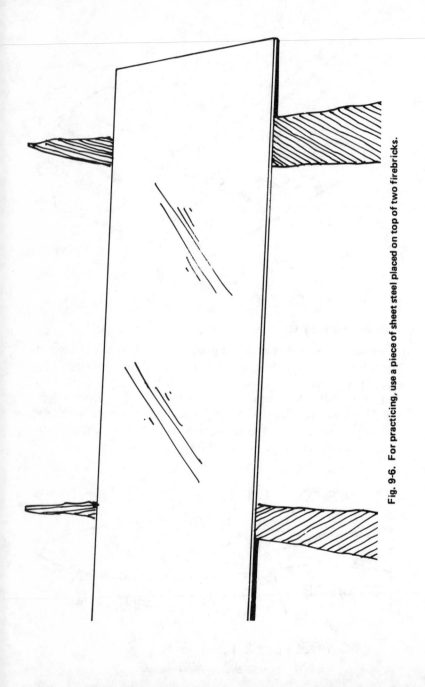

Fig. 9-6. For practicing, use a piece of sheet steel placed on top of two firebricks.

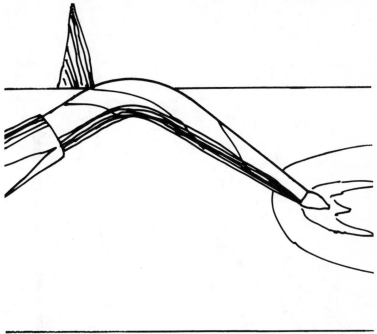

Fig. 9-7. Begin heating the steel in the middle and watch the metal change color as it gets hotter.

Heating Steel

Begin heating the steel in about the middle of the piece (Fig. 9-7). You should hold the torch in a comfortable and balanced position. The tip should be slanted at about a 45 degree angle to the right (southpaws should make suitable adjustments). The direction of travel will be in a circular motion from right to left. The circles you create should be about 1/4 inch in diameter. The tip of the cone of the flame should be 1/16 to 1/8 inch above the metal.

Watch the metal closely as it begins to heat up. It will turn cherry red, orange, yellow and finally white. About the time the metal turns white-hot it is in a molten state. While heating the metal up to this state, it is important that you keep the flame cone moving over an area of about 1/4 inch. If you simply keep the flame in one spot, that spot will turn white and finally become a hole (Fig. 9-8). Since this is only practice, you should experiment

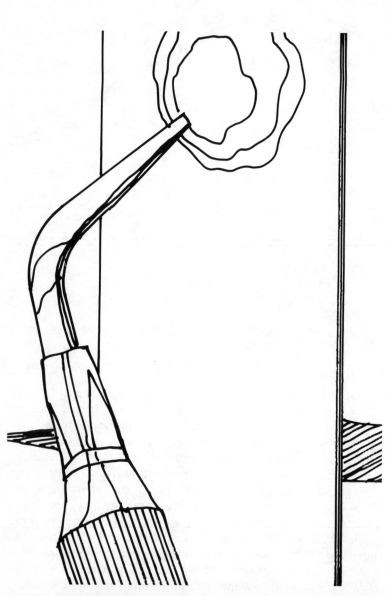

Fig. 9-8. Hold the flame in one spot until you have burned a hole in the steel.

with both holding the torch in a fixed position, which will cause a hole in the metal, and moving the torch in tiny circles causing the molten metal to puddle.

When the metal is in the molten state, you can move it around slowly with the cone in the flame. As you move the torch tip, you will notice that the molten puddle can also be moved. This, in fact, is one of the basic welding procedures. Much of your later welding will be carried out in a similar manner. In some cases, you will be using a filler rod, while in other instances you will simply be welding with the existing metal as we have just done in this puddling exercise.

Puddle Characteristics

There are a number of things that can be learned by carefully watching the puddle of molten metal. For example, the diameter of the puddle will be in proportion to its depth. On thin metal the puddle will or can be larger than on thicker metal. In the latter case, you must keep in mind that the heat necessary to create molten metal has to penetrate through the thicker metal, which is a large part of the reason for the smaller puddle. Thicker metal also requires a larger tip size to help create more heat.

The general characteristics of the puddle will give you a very good indication of the type of flame you are working with. A neutral flame will produce a puddle which appears smooth and glossy (assuming of course that the grade of metal is a good one, with few impurities or dirt). If the puddle sparks or bubbles excessively, this is usually an indication of either an improperly adjusted flame or that the metal is dirty or of a poor quality. The puddle will appear dull and dirty if the flame is more carburizing than neutral.

When a properly adjusted neutral flame is being used to create a puddle of molten metal, you will notice a small bright incandescent spot on the edge of the puddle opposite the flame. This spot will move quite actively around the far edge of the puddle. If the flame is not neutral, the spot will be oversized and will barely move.

Once you have learned how to develop a puddle of molten metal and have burned a few holes in this scrap piece of metal, the next step is to move the puddle over the surface of the metal. A brief description of the technique used will be of some help.

Welding Beads

The torch tip, with a neutral flame, is 1/16 to 1/8 inch above the surface of the metal and held at about a 45 degree angle to the

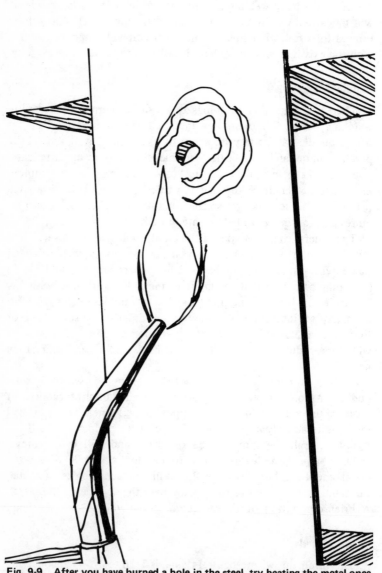

Fig. 9-9. After you have burned a hole in the steel, try heating the metal once again in another spot. Try to puddle the metal in a straight line.

surface. The direction of travel will be from right to left and in a circular motion (Fig. 9-9).

Begin forming a puddle just in from one edge of the sheet steel. As the metal begins to become molten, slowly advance towards the far end of the piece in a circular motion. As you move forward, the metal will go from a solid to a molten state. Then, as you move towards the opposite edge, the steel will become solid once again. You will notice that the circular movement of the torch will produce circular ridges as the metal cools. This line of ridges is commonly called a *welding bead*. Develop this welding bead across the face of the steel (Fig. 9-10).

With a little practice you should be able to create a welding bead that is uniform and straight. In the beginning, you will undoubtedly burn holes in the metal and probably develop a bead which more than slightly resembles a snake. After a bit of practice, however, you should be able to create a welding bead that is both uniform and straight.

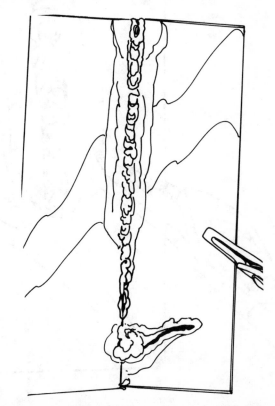

Fig. 9-10. A finished weld done without a welding rod.

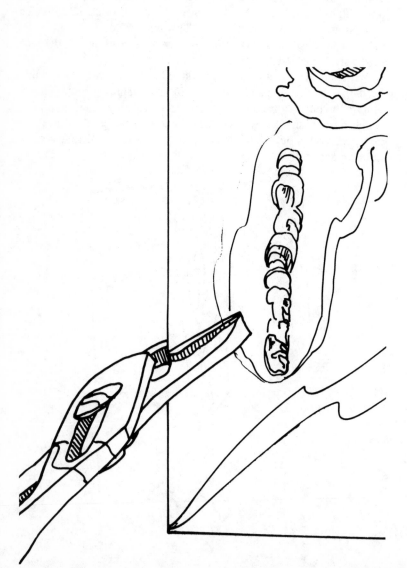

Fig. 9-11. Pick up the steel with a pair of pliers and examine the cooled puddle.

Penetration of a weld is very important. It is entirely possible to form a puddle of molten metal on the surface of the metal and not all the way through. When practicing, you should turn the scrap piece of steel over to see that the weld has penetrated through the metal.

As a side note, it makes a lot of sense to keep a pair of heavy duty pliers handy while welding. The pliers can be used for picking up the hot metal. This will safely enable you to inspect the weld (Figs. 9-11 and 9-12).

You should practice puddling sheet steel until you can create a straight, well penetrated bead welding across the face of the steel. Most knowledgeable welders will probably agree that if you can create five straight and well-formed welding beads, one after another, you are proficient enough to try more serious welding (Fig. 9-13).

Fig. 9-12.   Be sure to check the back side of the steel to see how well the puddle penetrated through.

Fig. 9-13. Practice puddling the steel with the torch until you can produce five straight and uniform puddles with good penetration.

WORKING WITH WELDING AND FILLER RODS

Until this point we have been practicing only with a single piece of steel and a flame. Since much welding uses a welding or filler rod as well, this will be our next objective. Since working with a welding torch and filler rod requires that you use both hands, the following exercise should not be attempted until you have reasonably mastered working with a torch to create a welding bead by puddling.

At this point, some mention should be made about welding or filler rods used by the oxyacetylene welder. Often during the oxyacetylene welding process, more metal than that which is melted by the torch is needed for a successful weld. To add more metal, it is common practice for the welder to use a welding rod which has the same, or very similar, properties as the base metal being joined.

Welding rods are available in many different alloys and designed for just about any type of joining task. Probably the most common type of welding rod is simply mild steel. These are commonly sold in 36-inch lengths and are copper-coated to protect the rod from rust and corrosion while in storage. Uncoated mild steel rods are also available. If there is an all-purpose welding rod, a mild steel rod is probably the one to buy. Diameters range from 1/16 to 1/4 inch. Your local welding supply house will have a selection of welding rods. You should buy an assortment of these mild steel, copper-coated rods. In most cases, you will use the 1/8-inch diameter rods with the greatest frequency (Fig. 9-14).

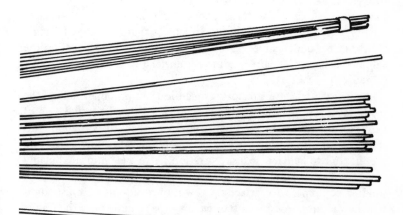

Fig. 9-14. Assortment of welding rods.

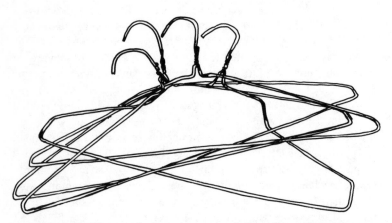

Fig. 9-15. Wire coat hangers are a poor substitute for welding rod.

Coat Hangers

Some mention should be made about the economy of using a substitute to the all-purpose welding rod. If you spend enough time in the company of welders, you will probably hear that a standard wire coat hanger can be used as a substitute for mild steel welding rod. While it may be true that a common coat hanger can be used as a filler material when welding, you should keep in mind that the steel used for these hangers is generally of a very low quality. It is therefore not wise to expect that a wire coat hanger will hold up under stress. An example may help to illustrate the value of using a standard welding rod instead of a wire coat hanger (Fig. 9-15).

An acquaintance was restoring an antique automobile. One of the many tasks which needed to be accomplished was the installation of new motor mounts in the engine compartment. It was about this time that the friend decided to purchase a new welding outfit and use the unit to help him in his restoration. After becoming reasonably competent at welding, he decided that it was time to start using the welding equipment for attaching new motor mounts. I must admit that he did an impressive job of welding the mounts onto the frame. The finished welds looked as if an old pro had done the work. The only problem with the project was that the welded mounts did not hold up under the torque of the engine because he used wire coat hangers instead of an all-purpose welding rod as a filler material. To make a long and rather expensive story short, my friend had to reweld the mounts. The second time around, however, he used standard copper-coated, steel welding rods and the welds have held to this day.

My friend is all that much wiser when it comes to welding. You can bet that he will never use wire coat hanger as a filler material ever again. In fact, in his entire household, you cannot find any type of clothes hanger other than a wood one.

Adding Filler Material

To produce a welding bead with the aid of a welding or filler rod, you begin in very much the same manner as when working solely with a torch. The major difference, of course, is that in addition to holding the torch in your right hand, you will also be holding (and moving) a welding rod in your left hand. The welding rod is used to add material to the puddle of molten metal. The more proficient you can become at adding filler material to a weld, the better your overall welding and the stronger your welds will be.

It is important that you become proficient at adding filler rod to a puddle of molten metal. To practice, begin in the same manner as the first exercise in this chapter, with a single piece of sheet steel suspended between two fire bricks. Hold the torch in your right hand and a welding rod in your left (Fig. 9-16). Begin heating the metal with small circular motions in about the middle of the piece of steel. As the metal begins to pass through the different color changes, slowly lower the tip of the welding rod into the flame to preheat it (Fig. 9-17). It is important that you do not touch the cone of the flame with the welding.

As the parent metal begins to become molten, the tip of the preheated welding rod is touched to the edge of the puddle. Almost immediately, the welded rod will melt into the puddle and mix with the parent metal (Fig. 9-18). After enough welding rod has been added to the puddle to create a slight crown on the surface, the tip of the welding rod should be withdrawn to about 1/8 inch above the surface of the welding bead.

Torch control is very important when introducing filler material to a weld. The flame should be in constant motion as should the tip of the welding rod. The cone of the flame does not cover a large area, but it is moving nevertheless. This will ensure even heat over a small area.

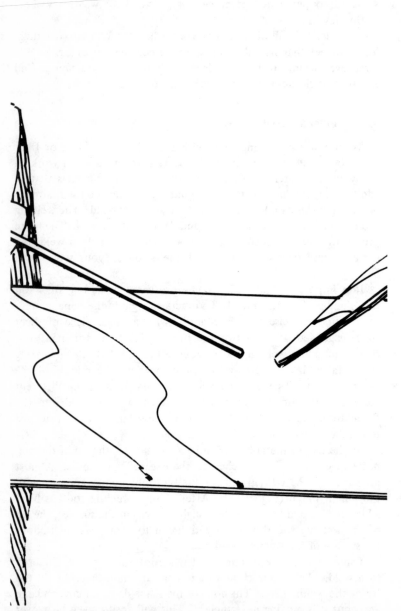

Fig. 9-16. Practice holding the torch and welding rod in the proper position before lighting the torch.

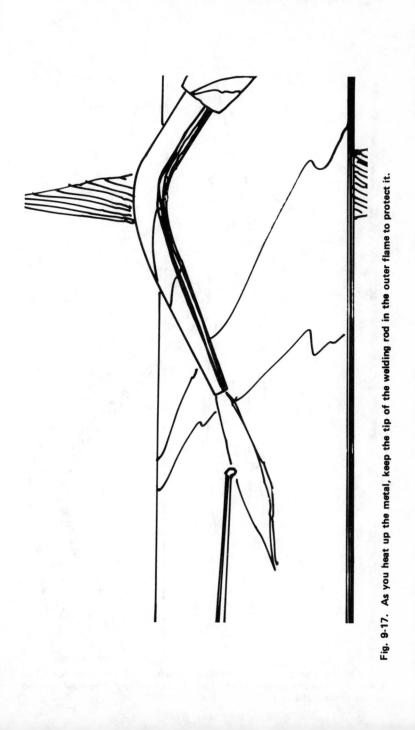

Fig. 9-17. As you heat up the metal, keep the tip of the welding rod in the outer flame to protect it.

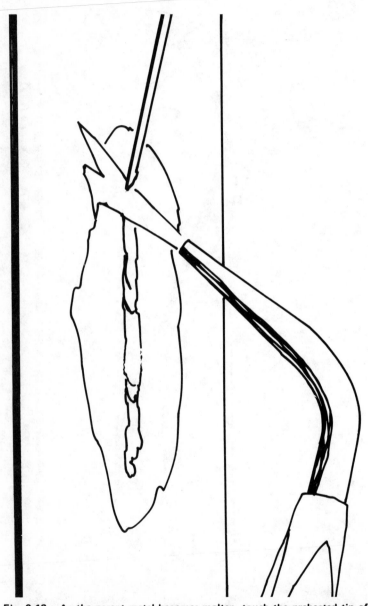

Fig. 9-18. As the parent metal becomes molten, touch the preheated tip of the filler rod to the far edge of the puddle. The rod will then flow onto the surface of the metal.

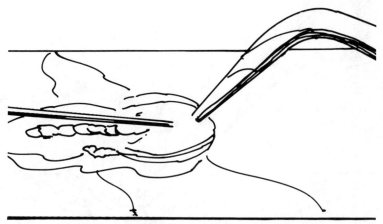

Fig. 9-19. A good weld bead produced with a filler rod.

Practice melting filler rod into a weld bead until you can produce five straight welding beads. Probably the most difficult technique to master when introducing filler rod into a welding bead is the circular motion of both hands. This technique can only be mastered through practice. It is extremely important that you practice adding filler material to the molten surface of the parent metal until you can produce a respectable welding bead (Fig. 9-19).

PROFESSIONAL POINTERS

At this time a few professional pointers are in order. It is important that the tip of the welding rod be preheated before it is touched to the surface of the puddle of molten metal. A rod which is outside the flame envelope will be much cooler than permissible (Fig. 9-20). When it is touched to the puddle, it will cause the molten metal to cool. The end result will be a poor quality weld bead.

On the other hand, if the tip of the filler rod is held too close to the cone of the flame, it will melt and drip into the puddle. More

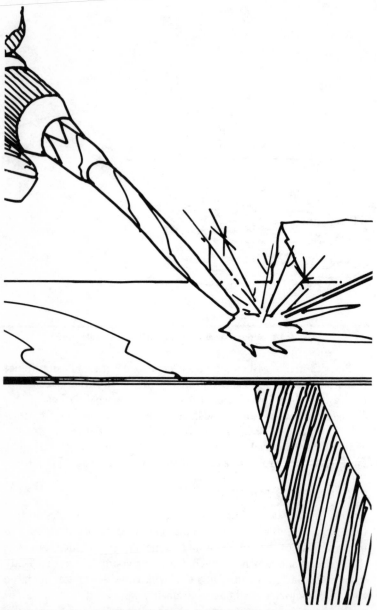

Fig. 9-20. It is important to keep the tip of the filler rod in the outside of the flame envelope.

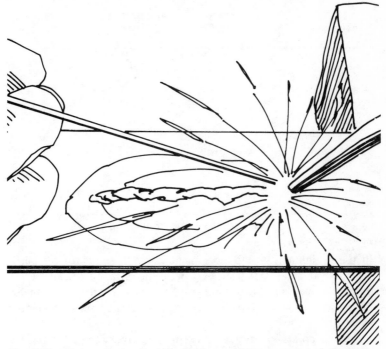

Fig. 9-21. The torch tip should be held at a 45 degree angle and the flame should be about 1/8 of an inch off the surface of the parent metal.

often than not this will result in the filler material being blown away from the weld which will produce an uneven bead and, more importantly, incomplete fusion.

Torch Control

Torch control when adding filler material is also quite important for a satisfactory weld. You should remember that the tip of the torch should be held at about a 45 degree angle to the surface of the parent metal (Fig. 9-21). When the torch is being held correctly with the proper flame, the surface of the puddle will be glossy and smooth. Often a slight angle adjustment is all that is required to produce a puddle with these characteristics.

Filler Rod Size

Another consideration for a successful and sound welding bead is the size of the filler rod. Table 9-1 offers some suggestions as to the diameter of the rod to use for the various thicknesses of metal.

METAL THICKNESS	WELDING ROD DIAMETER	PRESSURES OXYGEN	(PSI) ACETYLENE
1/16	1/16-3/32	4	4
1/8	3/32-1/8	5	5
1/4	5/32-3/16	8	8
3/8	3/16-1/4	9	9

Table 9-1. Suggested Welding Rod Sizes.

In the beginning, use the recommended welding rods for the particular job at hand. As you work with the different thicknesses of metal, you will develop a familiarity with how a particular diameter rod performs. You will therefore be able to create a suitable welding bead.

Some points to consider concerning improper diameter welding rods are worth mentioning at this time. If a 3/32-inch welding rod is the proper one to use for a particular joint and you switch to a thinner rod, say 1/16 inch, you will find it very difficult to add enough filler material to the weld bead. A smaller rod will also make controlling the puddle and resultant bead very difficult as well. The smaller rod will also tend to oxidize or burn very quickly.

Using a rod which is larger than what is called for can have similar effects on the finished weld. For example, if a heavy rod is used on light metal, it may create too large a puddle and therefore a sloppy weld bead. A heavier than required filler rod may also tend to cool the puddle of molten parent metal too much while it is being added. The end result will be poor penetration and an ineffective weld bead.

JOINING TWO PIECES OF STEEL

Once you can create a straight weld bead on the surface of a single piece of sheet steel, it is time to begin practicing joining two pieces of steel. For this exercise select two pieces of 1/8 inch thick steel strap, about 8 inches long. Lay the two pieces, with long edges aligned and butted together between two firebricks (Fig. 9-22).

Fig. 9-22. Lay two pieces of light steel on top of two firebricks. The edges should be aligned and as close as possible.

In order to hold the two pieces together while welding, the ends should be tack welded. *Tack welding*, as the name implies, simply means to spot weld one very small section of the pieces being joined. In addition to being a good way to learn how to heat up two pieces of metal to the melting point, tack welding is also a very useful means of holding two pieces of metal together while they are being welded along their length.

Tack Welding

Hold the lighted torch so that it balances well in your right hand. Move the flame close to the end of the two strips, that area where the two strips butt together. Move the cone of the flame slightly in a tiny circular motion over the joint so as to cover an area of about 1/8 to 1/4 inch. Watch the metal change from its normal color to red and finally to white-hot.

Just about the time that the metal turns white in color, it should begin to flow together. As soon as this happens, you must move the flame away from the joint. It is important to keep in mind that

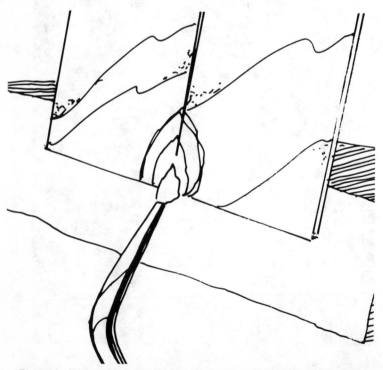

Fig. 9-23. Tack weld the joint edges of the two pieces without a filler rod.

if heat continues to be applied to white-hot and flowing metal, you will often cause a hole to appear. This is undesirable. Your aim is to simply bring the edges of the two pieces of metal up to the melting point, and then just a bit more so the edges fuse together and become one. Circular torch movement usually helps in fusing the pieces (Fig. 9-23).

Once a tack weld has been completed on one end of the strips, move the flame along the joint between the pieces and back a few times to heat up the metal slightly. Then tack weld the opposite end of the two pieces. Make sure that the pieces remain aligned while doing this tack welding.

In case you are wondering why the metal is heated up after one tack weld has been completed, this is so the two pieces will cool at approximately the same, even rate. This even cooling will reduce the chances of the strips buckling or warping as they cool.

Prefitting Pieces

There is nothing difficult about tack welding. But there are a number of procedures used by professional welders to insure the success of the weld. To increase your chances of producing a sound tack weld, you should know that the pieces to be tacked must fit tightly together along their entire length. Prefitting will make heating the two edges easier because you will not have to move the torch flame over a very large area. It will also enable you to flow the two metals a bit easier. If the two edges are not aligned snugly, you can still produce a strong tack weld. But you will find it necessary to use a bit of filler rod to make up the difference between the pieces.

Welding Irregular Edge Pieces

In some cases, it may be necessary to make a tack weld with the addition of a filler rod. One example is when the two pieces being joined have irregular edges. To accomplish this, begin by aligning the edges of the two pieces as closely as possible. Working on one end of the piece, heat both edges up to the white-hot stage and add the filler rod. As you work, you will find it necessary to keep the flame moving constantly. Add the filler material at the exact moment the metal begins to flow. If you wait too long, the metal from either or both pieces may begin to drip, necessitating adding even more filler rod.

After the second tack weld has been completed, the strips will remain in position while you weld the long joint between the two

Fig. 9-24. Puddle the edges of the joint without the aid of a filler rod. Keep the torch flame moving in tiny circles as the metal becomes molten.

pieces. Your first attempt at welding two pieces of metal together should be done without the aid of a filler rod. For the record, when tack welding long pieces of metal more than 12 inches in length, the edges are tack welded about every 6 inches.

Heat Application

Begin at one end of the joint between the strips. With small circular movements of the torch, heat up the edges of the joint until they are in a molten state and are fusing together. As the joint begins to fuse, slowly move the flame toward the far end of the joint to insure that complete fusion is taking place. It is important to keep the cone of the flame moving at all times, while at the same time working towards the far end of the joint. When working with thin metal, such as 1/8 inch thick or less, you will find that you can move relatively quickly along the joint (Fig. 9-24). With heavier metal, you will find it necessary to work more slowly and with a larger tip size.

In principle, this exercise in welding is the same as the first welding exercise in this chapter on puddling metal. The major difference, in this case, is that you are puddling the edges of two pieces of metal rather than simply working on a single piece of steel.

As you progress along the joint, it is important that you apply enough heat to fuse the edges of the pieces all the way through. Your best indication of the penetration of the weld will be the puddle being created by the cone of the flame. It will be helpful to you, especially at this early stage of metal joining, to stop welding and turn the pieces over to make certain that the weld has fused the edges of the pieces all the way through. Use a pair of pliers or a gloved hand for turning the metal over. If a circular weld appears on the back side of the joint, you are applying enough heat. However, if the weld does not show on the other side, you are not applying enough heat to the joint (Fig. 9-25).

WELDING JOINTS

Practice welding the joint between two pieces of light steel until you can accomplish this task with some competence. See if you can produce a strong weld without the aid of a filler rod. After you have mastered welding two pieces of light steel, where the butting edges are on the same plane, you should turn your attention to welding other types of joints, still without the addition of a filler material.

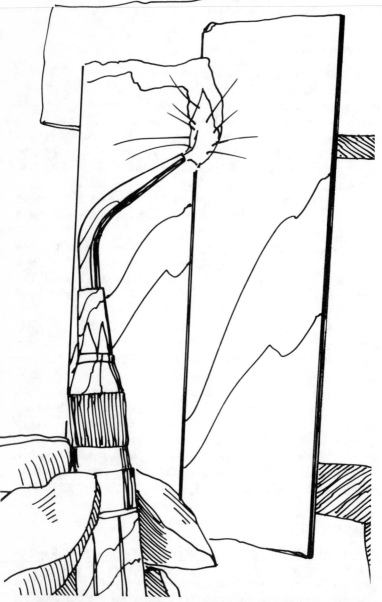

Fig. 9-25. After you have made one puddling pass along the joint without a filler rod, turn the joined pieces over with a pair of pliers and check the penetration of the weld.

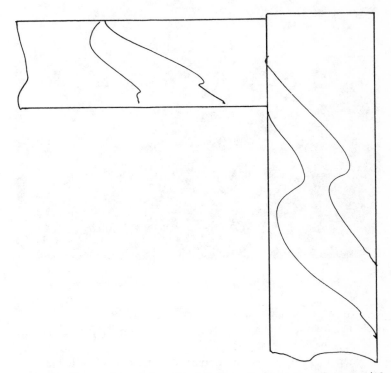

Fig. 9-26. An outside corner weld with the vertical member extending 1/16 to 1/8 of an inch above the horizontal member.

Outside Corner Weld

Another type of welded joint that does not necessarily require the addition of a filler rod is called the *outside corner* weld. For this exercise select two pieces of light steel, 1/8 inch thick, 8 inches long and 2 to 4 inches wide. Place them together at right angles so that the vertical piece extends beyond the horizontal piece by about 1/16 inch (Fig. 9-26). Begin by tack welding both ends of the strips to hold them in position while you weld the joint. The metal which extends will be used as a filler material as you weld.

A few professional pointers will help you to achieve success when making the outside corner weld. Begin by holding the torch so that the flame plays on the horizontal piece. This will cause the weld

Fig. 9-27. Proper torch position when welding an outside corner joint.

to form on the horizontal surface where it belongs. Cleaning up the weld bead will be much easier later on as you will only have to grind one side, the horizontal piece.

When making an outside corner joint weld, you will find that less torch movement is necessary than usual, particularly on butt joints (Fig. 9-27). You still must move the cone of the flame so that it heats up both pieces, but your main intention will be to "flow" the extended vertical piece onto the horizontal piece. This generally means less torch movement. You will also find that working from the outside of the vertical piece is the best way to approach this type of weld.

After the outside corner weld has been completed and the metal has cooled and solidified, check the weld first for appearance and then for strength. A careful look over the weld bead will give you a good indication of how sound the weld is. Next, check the strength of the weld by trying to bend the two pieces of metal, much the same as you would open a book (Fig. 9-28). If you can hear any cracking or breaking of the metal, this indicates that the weld did not penetrate very deep into the joint between the two pieces.

Fig. 9-28. Test the strength of an outside corner joint by trying to bend the pieces, much the same as you would open a book.

One of the unique features of the outside corner welded joint is that the weld need not be visible on the inside of the joints. In fact, the weld should be strong enough without penetrating all the way through. The best way to discover if the weld is, in fact, strong enough is to try to bend the two pieces apart. If the pieces hold without giving in the slightest, the outside corner weld is a good one.

Flange Joint Weld

Still another type of weld that can successfully be made without the aid of a filler rod is called a *flange joint weld*. To make this type of joint, you must first bend both edges of the pieces to be welded so that each piece will have an edge with a 90 degree angle.

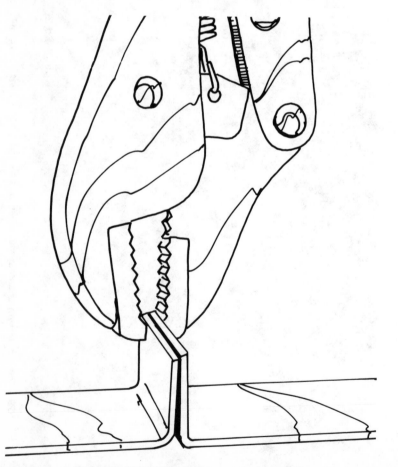

Fig. 9-29. Flange joint held by a pair of vise-grips.

The flange should be from 1/4 to 1/2 inch wide and should be roughly equal to the piece it will butt against (Fig. 9-29). A heavy duty bench vise is quite handy for bending the edges of sheet steel.

Begin a flange welded joint by aligning the flanges of the pieces and tack welding both ends to hold the pieces securely while welding. Next, heat from the torch is applied to the top of the flanges until they melt and fuse together. As the two flanges melt together, enough filler material from the parent metals should be present to produce a strong and sound weld (Fig. 9-30).

As you can see, there are a number of welds that can be accomplished without the addition of a filler material. Generally speaking, this is the case more often than not when working with thin steel. As the thickness of the steel increases, however, you will find it more difficult to develop a built-up welding bead. In fact, at this time you may have discovered that the bead you can produce by simply puddling the metal is slightly indented. Whenever this happens, it means that additional filler material is required. For our purposes, that will mean welding rod. It will probably be most helpful if I describe how to make some of the more simple welds using a filler material on relatively thin steel. Keep in mind, when adding filler material in the form of welding rod, that we are approaching the joining process in very much the same manner as when joining metal without a filler rod, that is, puddling the metal. The major difference, of course, is that now we are building up the weld bead with the filler material. Basically, everything else is the same.

Butt Joint Weld

Since the butt joint is the easiest of all joints to accomplish, you should begin welding with filler rod with this simple arrangement. Lay the two pieces of sheet steel, about 1/8 inch thick, side by side and suspended between two firebricks. Next, tack weld both ends of the strips to hold them in position while you weld the joint. Begin, if you are right-handed, on the right end of the two pieces and work towards the left. For welding metal less than 1/4 inch thick, use the forehand welding method described earlier in this chapter. To briefly refresh your memory, the forehand method of welding has the welding rod held in position over the as yet unwelded section of the joint. The flame of the torch is held at about a 45 degree angle to the surface of the joint. The flame is played on the joint in a circular motion, while the tip of the welding rod is held inside the flame envelope to preheat it. As soon as the metal edges of the joint become molten and begin to flow together,

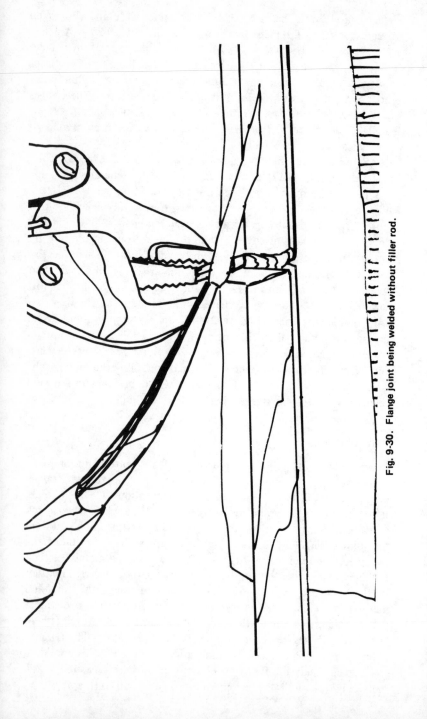

Fig. 9-30. Flange joint being welded without filler rod.

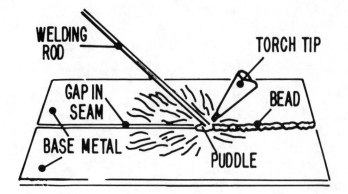

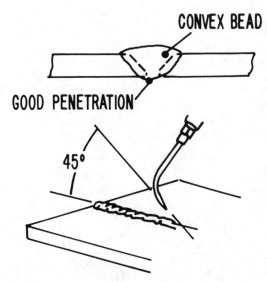

Fig. 9-31. The filler rod is applied as the weld metal becomes molten and starts to flow.

the tip of the welding rod is touched to the far edge of the puddle. This will cause some of the filler rod to flow into the joint and become part of the weld bead (Fig. 9-31).

You must move smoothly along the seam, constantly keeping the flame moving to evenly heat both sides of the joint. In the beginning you may find it difficult to hold the torch at a constant height above the joint, while at the same time holding the tip of the welding rod in the flame envelope. If you have a problem coordinating both hands, try locking your elbows against your sides and keep your forearms at a 90 degree angle to the front of your body. In

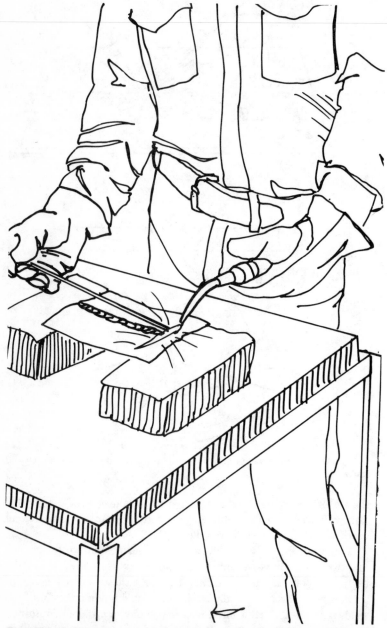

Fig. 9-32.   Try locking your elbows against your sides to steady both the rod and blowpipe actions.

time you will develop this technique. When you are first starting it seems difficult, hedging on the impossible (Fig. 9-32).

As you move along the joint, it is important to keep the flame moving at an equal distance over both edges. You will not be covering a very long span at any one time. Instead, make sure that you are creating a good puddle of molten metal. When the edges of the joint begin to fuse together, lower the tip of the preheated welding rod into the puddle. Allow the filler rod to flow into the puddle until the height of the welding bead is built up slightly. Then proceed a little bit further along the joint and repeat the heating. Your aim, when adding the welding rod, is to build up a joint that is equal along its entire length.

Whenever you come upon a tack weld, heat the area as if it did not exist, and add the filler rod. During the entire weld, the filler rod should be added in uniform amounts and at regular intervals. It is extremely important that you do not try to rush through the welding. Allow the edges of the metal to heat up sufficiently and to flow so good penetration and fusion can take place.

Practice the butt weld, with the addition of a filler rod, until you can produce a respectable weld bead with good penetration through the joint. Work only with sheet steel that is less than 1/4 inch thick until you have perfected and mastered the torch and rod movements. After you can make a decent butt weld, the next exercise should be to learn how to make a T-joint.

T-Joint Weld

T-joints are formed by placing one sheet of steel on a similar piece of steel so that the two pieces resemble Fig. 9-33. The next step is to tack weld the ends to hold the pieces in position while you weld the joints. After the tack welds have cooled for a few minutes, the next step is to weld the vertical piece to the horizontal piece. The forehand method of welding should be used initially. Once you have reasonably mastered this technique, try the backhand method.

When welding a T-joint, it is important that you do not allow too much heat to build up on the vertical piece or you may burn through it. While the flame of the torch should be kept in motion during this welding, you will probably find that the pattern of movement will be very tight. Lower the tip of the welding rod into the joint when the metal of both pieces starts to flow. It is important at this time to keep an eye on the vertical piece as it will always have a tendency to reach a molten state quicker than the horizontal piece.

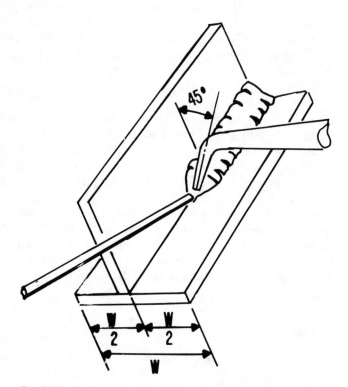

Fig. 9-33. Welding a T-joint using the forehand welding method.

If you find this to be a problem, adjust the angle of the flame so it is closer to vertical than horizontal (Fig. 9-34).

After the joint has been welded uniformly along its length, allow the metal to cool and then inspect the weld. The bead should be built up to an equal height along its length and be equally distributed along both pieces. In other words, one side should not have more of a bead than the other. The vertical piece should appear to be solid and the same thickness as when you began. If the vertical piece looks thinner in certain areas, this is usually an indication that you have melted some of the metal away. The thinner area would not have as much strength as the rest of the joint (Fig. 9-35).

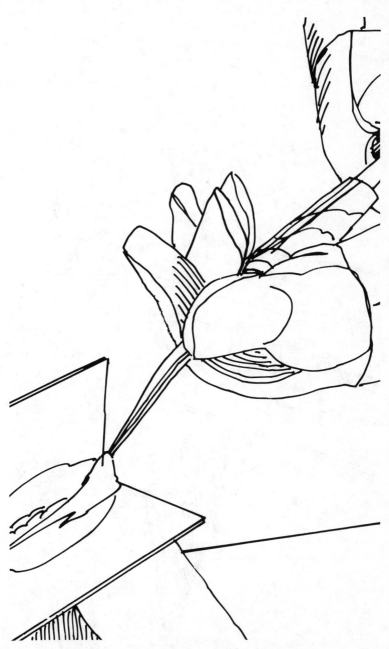

Fig. 9-34. Adjust the torch angle more towards 90 degrees and you will have better control when making a T-joint.

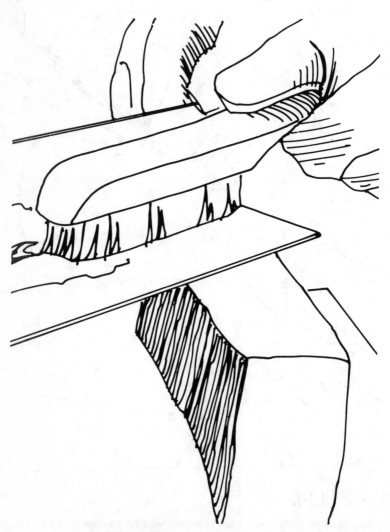

Fig. 9-35. Clean up the joint with a wire brush.

Chapter 10
Welding Thicker Metals

We have covered how to weld relatively thin pieces of steel both without and with the addition of a filler material. In just about every instance, thinner metal does not require any edge preparation other than making sure that the joint fits together well and is clean. When working with thicker metals, however, you will generally not be able to achieve good penetration unless you reduce the thickness of the metal along the joint area. It is generally accepted that when welding any metal that exceeds 1/4 inch in thickness, you must reduce the thickness of the joint edges by grinding. Then fill in the area with welding rod. Actually, the point of edge treatment is to insure that the heat from the oxyacetylene flame can be total along the joint. The thicker the metal along the edges, the more heat must be applied to make the edge metal molten throughout. As it turns out, metal thicker than 1/4 inch cannot be heated thoroughly enough to make fusion complete and total. This is the basis for edge treatment.

Basically there are four types of joints in all welding: butt, T-joint or fillet, flange and lap. Since we have covered these previously, the information will not be duplicated here (Fig. 10-1).

BEVELING METAL EDGES

When working with thick metal, some type of edge treatment must be initiated before welding begins. In many cases this will simply amount to beveling one or more of the edges at about a 45 degree angle. Then, when the two pieces are placed together prior to welding, the beveled edges will form a "V" which can then be welded. In some cases the depth of the "V" groove will be deep, requiring more than one pass to be made with torch and welding rod.

There are really only a few ways in which the edge of a piece of metal can be beveled. These include *grinding, cutting*

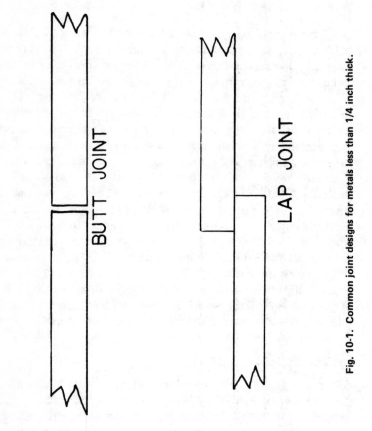

Fig. 10-1. Common joint designs for metals less than 1/4 inch thick.

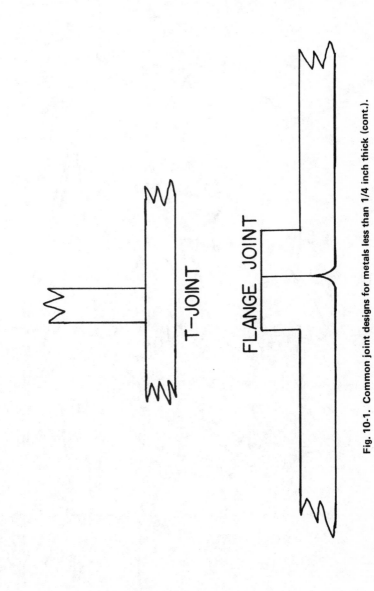

Fig. 10-1. Common joint designs for metals less than 1/4 inch thick (cont.).

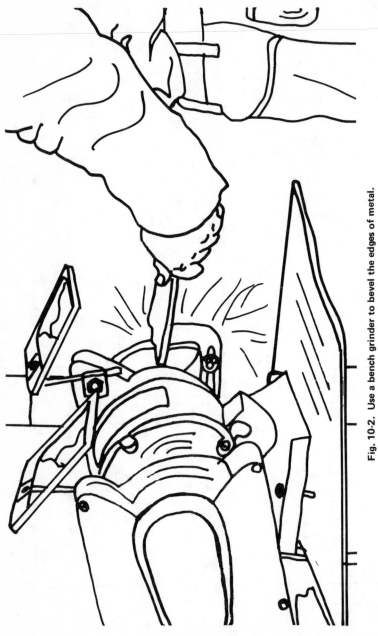

Fig. 10-2. Use a bench grinder to bevel the edges of metal.

and *machining*. In most cases only the first two are within the means and capabilities of the average home welder, machining being a commercial production form of edge preparation.

Grinding

Grinding the edges of metal is probably the easiest and most efficient means of edge beveling for the do-it-yourself welder to perform. Since every home workshop should have some type of grinding wheel, chances are good that you will already possess the equipment necessary for this special edge preparation. Begin by marking the angle of the bevel, in most cases 45 degrees, and press the metal into the spinning grinding wheel. For larger pieces, you may find the work easier to accomplish if you set up some type of guide on the front of the grinding wheel (Fig. 10-2).

One alternative to the conventional grinding wheel is a disc sander fitted with a special metal cutting abrasive wheel or blade. This hand-held machine can do double duty. It is handy for cleaning up metal before and after welding prior to finishing the metal with paint (Fig. 10-3).

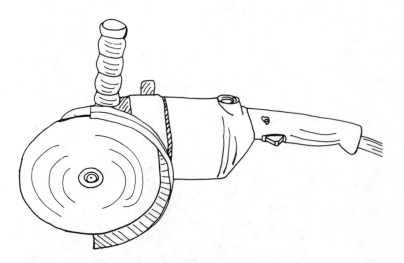

Fig. 10-3. A hand-held disc sander is very useful for cleaning up metal before and after welding.

Fig. 10-4. Beveled edges can also be made with a cutting torch. The whole key is in holding the torch at the right angle and steady.

Cutting

Cutting with an oxyacetylene cutting torch is another way the do-it-yourselfer can bevel the edges of metal prior to welding (Fig. 10-4). This beveling can often be done while the metal is being cut to size. It is important to keep in mind, however, that the cut must be of a high quality to minimize any additional metal edge preparation and to insure that the pieces fit well together. It is generally accepted that cutting a beveled or chamfered edge on a piece of metal is not for beginners. You will therefore be better off to rely on a grinding wheel for your beveling work until you have become quite proficient with both welding and cutting.

WELDING STEEL PLATE

For your first exercise in welding thick metal, select two pieces of mild steel plate, 6 to 8 inches long and about 4 inches wide. Steel plate, in case you are wondering, is metal that is at least 1/4 inch thick. For this first exercise, you should choose plate that is about 1/4 inch thick. Steel of this thickness will be more than adequate for learning how to weld heavy metal.

Before the welding can begin, you must first bevel the edges that will butt together. As mentioned earlier, the best way to do this is on a bench grinder. Any quality bench grinder will have a guide table in front of each wheel which can be set so that grinding will take place at a predetermined angle. Grind the one long edge of both pieces of steel plate to an angle of 45 degrees. It may be helpful at this time to explain how to grind the edge of a piece of steel plate.

Grinding Plate Edges

You should make sure that the bench grinder is securely fastened to a work surface. Good lighting should be over the machine, and you should have on clear eye protection. Lightly press one end of the plate into the spinning grinding wheel and move the piece with a back and forth motion. The reason for this is so you will not grind one area more than another. Do not press the metal too hard into the wheel as this will slow down the spinning and reduce the speed at which the metal is removed. During the grinding, it is important that you keep the steel plate flat on the guide in front of the machine (Fig. 10-5). This will insure that the angle being ground is, in fact, consistent. Stop grinding often and carefully inspect the edge to make sure it is being ground at both the right

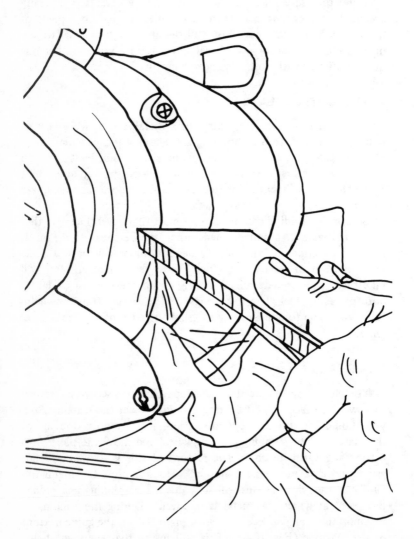

Fig. 10-5. Use a guide to help you grind a true bevel on steel plate.

angle and uniformly along the edge. Proper grinding takes time, so do not try to rush the work.

Tack Welding Joint Edges

After both edges have been ground, lay the two pieces of steel plate across two firebricks, with the beveled edges aligned. Tack weld both ends of the joint to hold them in alignment while you weld the beveled joint (Fig. 10-6). Use a bit of filler material if you find it necessary for the tack weld.

As a side note, you should have already attached the proper size welding tip into the blowtorch blowpipe handle. When in doubt as to the proper size welding tip, consult the handbook which came with your torch. Here you should also find recommended working pressures for both oxygen and acetylene. See Table 10-1.

Begin welding the joint at the left end, heating the edges of the joint thoroughly before introducing the welding rod. For 1/4-inch thick steel plate, a 3/32 or 1/8-inch thick all-purpose welding rod is adequate. It is important, when working with thicker metals, to keep the torch flame moving to produce thorough penetration. The size of the molten puddle will be the largest you have ever created, so give the metal time to heat up and flow.

Once the puddle is formed at the bottom of the V-groove, introduce the tip of the preheated welding rod. Flow the rod into the groove and build up the metal until it has filled the groove and risen to about 1/16 inch above the joint. Continue along the joint in this manner until you have both filled the groove and built up the weld bead.

Basically, as previously stated, welding steel plate is quite similar to welding sheet steel. The major difference, of course, is that you must generally work slower and add more filler material to produce a strong weld. By beveling the edges of the joint before welding, you are almost guaranteeing good penetration of the weld. One of the major points to keep in mind is that you must evenly heat up both sides of the joint to the molten stage before adding the filler material rod.

Strength Test

After you have welded your first joint in steel plate, you should check the weld for strength. One very good way to do this is to first let the metal cool. Then clamp it in a vise and break the joint with

Fig. 10-6. Tack welding the ends of a beveled edged joint.

Table 10-1. Pressure Chart for Sears Welding Tips.

METAL THICKNESS	TIP SIZE	WELDING ROD DIAMETER	OXYGEN PSI	ACETYLENE PSI
1/32	1	1/16"	5	5
3/64	2	1/16	5	5
1/16	3	1/16	5	5
3/32	4	3/32	5	5
1/8	5	3/32	5	5
3/16	6	3/32	6	6
1/4	7	1/8	7	7
5/16	8	5/32	8	8

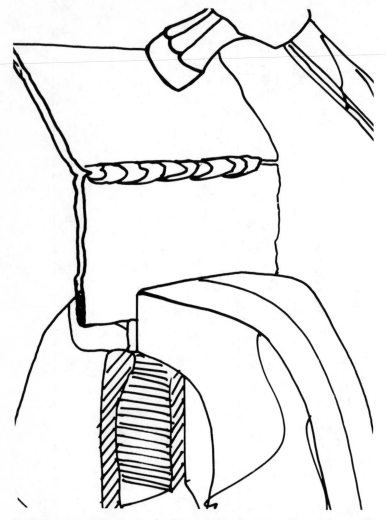

Fig. 10-7. Test the strength of a joint by breaking it in a vise. A good weld will require a number of hammer blows while a weak joint will break quite easily.

repeated blows from a heavy hammer (Fig. 10-7). A sound weld will require a lot of hammering to break while a poor weld will break quite easily. Practice the above welding exercise until you can produce a welded joint that is uniform and strong (Fig. 10-8).

Characteristics of a Good Weld

It may be helpful at this time to describe what a good weld looks like when working with steel plate. The weld bead should be smooth

Fig. 10-8. Practice until you can produce a series of straight and uniform welding beads.

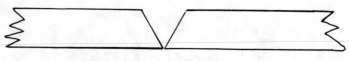

BEVELED BUTT JOINT

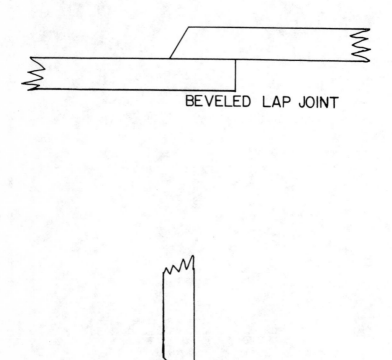

BEVELED LAP JOINT

BEVELED T-JOINT

Fig. 10-9. Common joint designs for metal more than 1/4 inch thick.

and form an equal junction between the two pieces of steel plate. The width of the bead should be uniform and about 25 percent wider than the metal is thick. For example, if the steel plate is 1/4 inch thick, the width of the weld bead should be about 5/16 inch. The underside of the weld should be uniform as well with no excess metal formations. In addition, the bead should be covered with a very thin oxide film on both sides of the joint.

Figure 10-9 shows most of the joint possibilities when welding thick metal. Since these are time-proven joint designs, you should use them whenever possible. All of these joints can be fashioned on a conventional home bench grinder, so the do-it-yourselfer should have little problem in duplicating them.

MULTI-LAYER WELDING

On very thick metal, over 5/8 inch, the home welder may find it almost impossible to fill the beveled joint with one pass of the torch. It is recommended that the welder build up the weld bead in several passes. Technically, this is called *multi-layer welding*. By approaching heavy metal welding in this manner, the welder will

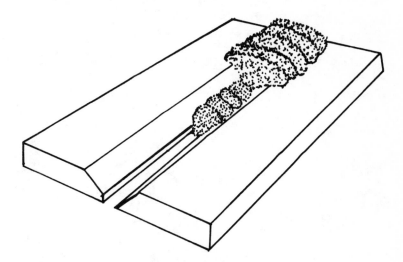

Fig. 10-10. Multi-layer welding.

find it much easier to deposit several layers of filler material until the welded joint has been built up to the proper height above the base metal. The point of multi-layer welding is to produce smaller puddles of molten metal and enable the welder to have more control over the joint.

When building up a weld bead with multi-layer welding, the first layer should provide good penetration at the bottom of the V-groove. Additional layers must fuse this first layer with the filler material and the sidewalls of the joint. The final layer seals the joint and should be crowned slightly above the base metal (Fig. 10-10).

Anyone knowledgeable in the art of welding will tell you that you should learn welding in gradual stages, mastering each before moving on. For the home welder, it would be best to begin with thin sheet steel. After mastering welding with and without filler rod, move onto steel plate. Do not attempt to weld plate thicker than 1/4 inch until you are able to produce a respectable welded joint. After you have the basics of torch control down, then move onto thicker metal.

Chapter 11
Protection and Problems

No welding project can be considered finished until the piece has been pressed into service. In some cases this will mean simply cleaning up the weld to remove the ridges of the weld bead. In other cases, however, this may mean grinding the weld flush with surrounding areas and finishing with several coats of paint or clear finish. This chapter explains protection techniques and examines some common welding problems.

PAINT COATINGS

The actual use of the welded piece will determine the extent of the finishing. You should keep in mind that ferrous metals (iron, steel and alloys) will rust if subjected to the forces of the elements. Since rust will in time cause the metal to deteriorate, you will want to prevent this whenever possible.

There are a number of paint coatings that have been specifically developed for use over metal surfaces. Generally speaking you cannot cover metal with just any paint, such as latex wall paint, and expect it to perform in service. You should also keep in mind that the first coat is usually a special primer covering designed to provide a suitable adhesion base for the subsequent finish coats. Probably the best place to find more information about paint coatings is at your local paint store.

One alternative to a paint coating for welded metals is some type of oil finish. Simply stated, if you coat metals which will be exposed to the elements with oil or some type of silicone spray (such as WD-40), the coating will reduce the chances of the metal rusting. If you do this, keep in mind that the coating must be removed before the metal is ever again heated with a torch flame. Oil or grease burns violently in the presence of pure oxygen, a very dangerous condition to say the least.

SANDBLASTING

Before any coating is put on metal, the surface must be cleaned of any existing rust, flux or scale. Some of the tools that will remove these surface coatings are a wire brush (hand-operated or mechanical), sandpaper, grinding wheel or other suitable machine. Sandblasting is still another way of cleaning up a metal surface preparatory to finishing. Since sandblasting equipment is expensive, it is recommended that the do-it-yourself welder have this cleaning work done by someone who specializes in the field. A good sandblast cleaning will give a truly professional look to your metal work.

COMMON WELDING DIFFICULTIES

Undoubtedly the most common problem encountered by beginners is popping of the torch during the welding process. In most cases, this is a result of holding the cone of the flame too close to the molten metal. As you develop some skill at keeping the torch tip at a consistent height above the work you will find popping less of a problem.

Torch Popping

Another cause of popping, and one that often results in the flame going out, results when the welding tip becomes dirty or clogged from bits of molten metal. The first thing you should do when the flame pops out is to turn off the acetylene and oxygen, in that order. Next, check the tip to see that it is not clogged. If it is, clear it with a special tip cleaner just smaller than the opening (Fig. 11-1). You should also remove any accumulation of metal slag around the orifice at the end of the tip.

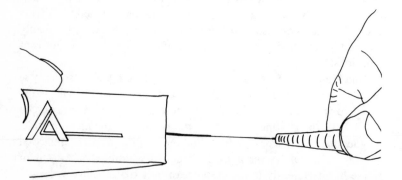

Fig. 11-1. Clean your welding tips with a special tip cleaning tool.

RANGE OF THICKNESS	NUMBER OF LAYER	METHOD OF DEPOSITING	CROSS SECTION OF WELD
3/8" TO 5/8"	2	TWO LAYERS ONE PASS	
5/8" TO 7/8"	3	THREE LAYERS TWO PASSES 1&2 IN ONE PASS LAYER 3 ONE PASS	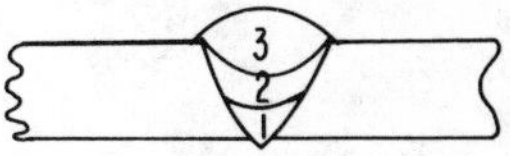
7/8" TO 1 1/8"	4	FOUR LAYERS THREE PASSES 1&2 IN ONE PASS LAYERS 3 AND 4 IN TWO PASSES	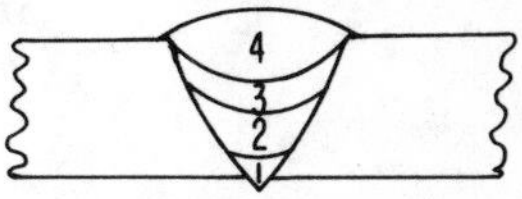

Fig. 11-2. Suggested methods for multi-layer welding.

If you move too quickly along a joint or if the oxygen/acetylene pressures are too low, the bead that will result will be narrow and uneven. When inspecting your practice welds and subsequent welds as well, keep an eye on the quality of the bead. A good weld will resemble Fig. 11-2.

Weld Bead Holes

The importance of constant torch movement cannot be over-stressed. Holes in a weld bead are caused by holding the cone of the flame in one spot for too long of a period (Fig. 11-3). If holes appear while the metal is still in the molten state, they can usually be filled with a bit of welding rod, properly applied.

Strive for perfection in your welding exercises. A good weld bead should be a uniform series of circles with no ridges or high and low spots. This uniformity can only come as a result of a total familiarity with your torch, its adjustment, the filler material and the metal being joined. Needless to say, it will take you many hours of working with a torch and the various types of metals before you can expect to produce good looking and sound welded joints.

Warping of Metal

Metal has a tendency to warp when heat is applied. This is a problem that is more prevalent with light metal; in short, the thinner the metal, the greater the possibility of some type of warpage. Knowing that metal will expand when heated and contract when cooled can be used to your advantage when working with relatively

Fig. 11-3. Holes in a weld bead are caused by not moving the blowpipe quick enough.

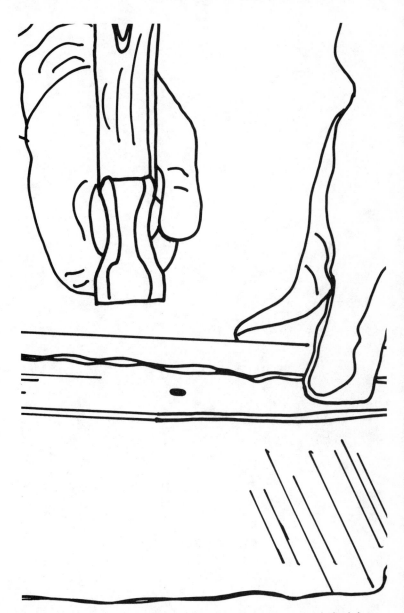

Fig. 11-4. Warped metal can often be hammered back into the desired shape.

thin sheet metal. After welding, straightening can be done with your torch. If undesirable warpage presents itself, you can usually correct the problem by hammering and/or heating the opposite side of the warp (Fig. 11-4).

Chapter 12
The Weldability of Certain Metals

You already know from reading Chapter 5 about some of the general characteristics and properties of many metals and their alloys. What you probably do not know, at this point, is how easy or difficult it is to weld these different metals. The purpose of this chapter then, is to supplement the information in Chapter 5 and let you know what you are in for when you are faced with welding a specific metal.

IRON

Wrought iron is very easy to work with because it approaches the melting point gradually and is therefore easy to control. Wrought iron can be welded with all-purpose steel welding rods quite satisfactorily. There are also low carbon content steel welding rods that have been designed specifically for welding wrought iron. You should use this type whenever possible.

Since cast iron is a chromium alloy, it can be welded very much the same as mild steel, with a neutral flame. Actually, there are two types of cast iron, gray and white. The gray cast iron is the easier to work with. In just about every instance, it is better to braze weld. Square cast iron filler rod, rather than round, is also available for welding cast iron (Fig. 12-1).

STEEL

Steel, in one form or another, is probably the most common metal for the do-it-yourself welder. As you know, steel is classified according to its carbon content and/or its alloying element, such as chromium nickel, manganese or molybdenum.

Mild steel or low carbon steel is the easiest type for the home welder to work with. Low carbon steel is probably the best steel to use when working with the oxyacetylene process. Use mild steel, all-purpose welding rods for best results.

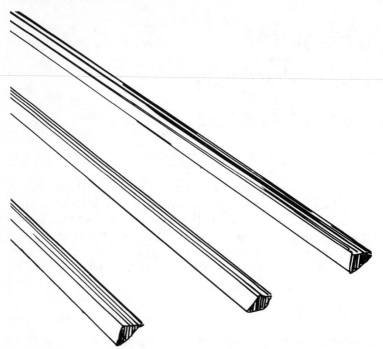

Fig. 12-1. Cast iron welding rod is required.

High carbon steel can be difficult to weld for the inexperienced hand because you must work quickly and keep the heat as low as practicable. All-purpose mild steel welding rod can be used for joints with high carbon steel. Stronger welds can only be achieved by using special high carbon filler rod.

Stainless steel is a poor choice for the oxyacetylene welder because it contains both nickel and chromium. The best method for welding stainless steel is TIG welding. If you find that you must weld stainless steel, keep in mind that it has a tendency to buckle and warp more than other types of steel. Reduce the chances of this becoming a problem by tack welding at close intervals before actually beginning the welding. You must weld stainless steel at an even, steady speed. One professional guide to remember is if the joint between the unwelded pieces begins to open as you weld, you are moving too slow. On the other hand, if the joint begins to close in front of the flame, you are moving too quickly.

Low alloy steels are generally quite easy to weld, especially when working with sheet type alloy steel. These steels generally have a low (0.15 or less) carbon content. An all-purpose mild steel filler rod works about the best.

Steel which contains a large portion of an alloying element requires that a special welding rod be used for a sound joint. Obviously, the filler rod should contain approximately the same proportion of the alloying element as the parent metal. Your local welding supply house is probably the best source of information about special welding rods.

COPPER

There are two types of copper: *electrolytic* and *deoxidized*. The former is difficult to weld because it contains a small amount of oxygen. It is usually brazed with a silver rod rather than welded. Deoxidized copper is easier to weld, providing a special deoxidized copper welding rod is used as a filler material.

Part of the difficulty in welding both types of copper is that the metal does not readily flow until it reaches the molten state. In other words, the metal remains solid until it reaches its melting point, about 1,980 degrees Fahrenheit. Once this temperature is reached, the copper will immediately become liquid.

Bronze, as you know, is an alloy of copper. This metal is very easy to weld and is quite popular for metal sculpture. When welding bronze, the obvious filler material to use is bronze brazing rod. You must also use a special flux when welding bronze. The flux forms a protective coating on both the metal and filler material which prevents oxidation of the parent metal during the meeting. The flux also dissolves the oxides which may form and further helps in the flowing of the filler material. Generally speaking, just a tiny bit more oxygen should be used in the flame when welding bronze. This will mean that the flame will be slightly oxidizing. Since the amount of excess oxygen will be influenced by the particular bronze you are working with, a bit of experimentation is necessary when welding this metal.

ALUMINUM

Aluminum is the most common metal in the world and is also difficult to weld with the oxyacetylene process. Part of the reason aluminum is hard to weld is that it has a low melting point, about 1,220 degrees Fahrenheit. The metal gives no warning as it approaches this temperature. A special welding process, MIG welding, was developed specifically for working with aluminum because of the problems of welding with the oxyacetylene joining process.

Aluminum, nevertheless, can be welded with oxyacetylene, but it is a chancy proposition. Begin by preheating the metal with an excess acetylene flame. This flame will deposit carbon on the surface and make it look very black. Next, heat the joint area with a neutral flame until the carbon is removed. At this point apply some special aluminum welding flux, which is liquid. Then, while keeping the cone of the flame at least 1 inch above the joint, introduce a special aluminum welding rod into the flame. The object here is to tin the joint with the aluminum rod. After the joint has been tinned, melt more rod onto the area and build up the joint. It is important that the flame of the torch not be played on the joint as this may cause the aluminum to become very weak. After the weld has been completed, all flux must be removed. This is commonly done with warm water.

There is no special talent required to learn how to weld. What is required, however, is a determination to master the basics of flame control and a great deal of patience. You should practice on lighter sheet steel in the beginning and not move onto thicker metals until you have mastered the basics of torch and flame control. In time you will develop a certain sense about how metals are affected by intense heat, how metals flow and just how much heat is necessary to manipulate molten metal. Practice until you know the characteristics of one metal; then expand your knowledge by experimenting with other types.

Chapter 13
Cutting Metal

Undoubtedly, the most dramatic use of oxyacetylene equipment is in cutting metal. Sparks shower from the work and, in most cases, the cutting progresses very quickly. The equipment necessary for cutting metal with the oxyacetylene process is basically the same as for welding. The major differences, of course, are that the blowpipe must be fitted with a special cutting attachment. The working pressures are generally greater than for standard welding (Fig. 13-1).

EQUIPMENT

In most cases, a special cutting attachment is fastened onto the end of the standard welding blowpipe handle in place of the welding tip. It will be to your advantage to understand the principles behind its operation (Fig. 13-2).

Cutting Attachments

Basically, all cutting attachments are the same, although design features often differ between manufacturers. A cutting attachment will have a housing through which three tubes run. One tube carries acetylene gas to the tip. Another similar sized tube will carry oxygen to the tip. A third, larger tube will also carry oxygen to the tip. The first two tubes carry oxygen or acetylene to a special expansion chamber just behind the cutting tip. Here the gases are allowed to expand and mix thoroughly before passing through the small holes in the cutting tip.

In most cases there will be six small holes in a cutting tip, surrounding a larger hole in the center. The third tube in the cutting attachment handle carries pure oxygen and it leads directly to the large center hole. The flow of the pure oxygen tube to the center hole is controlled and regulated by a lever which sits on top of the cutting attachment handle. No oxygen can flow through this tube until the lever is depressed (Fig. 13-3).

Fig. 13-1. Cutting in progress.

Fig. 13-2. A typical welding torch with cutting attachment.

Cutting Tip Holes

The six small holes in the cutting tip are called preheat holes and the larger, center hole is called the cutting hole. Each of the preheat holes produces a flame which resembles a standard welding flame. Since the mixed oxygen and acetylene flows equally through all of these small holes, each flame will be the same: acetylene, carburizing, neutral or oxidizing. As you can well imagine, the six preheat holes produce an enormous amount of heat (Fig. 13-4).

The center hole in a cutting torch tip does not produce any flame. Instead, when the lever is pressed, a stream of pure oxygen is introduced into the heat zone created by the preheat flame. This causes the metal to oxidize away very rapidly. It may be helpful at this time to explain how the oxyacetylene cutting process works.

CHEMICAL REACTION

Oxyacetylene flame cutting can be used for some ferrous metals. Oxyacetylene cutting can successfully be used to cut low carbon steel, low alloy steel, wrought iron and other ferrous metals. This process does not work well for cutting stainless steel, cast iron or any nonferrous metal such as aluminum.

Actually, the process of flame cutting is a chemical reaction. This reaction takes place because oxygen has a chemical attraction to

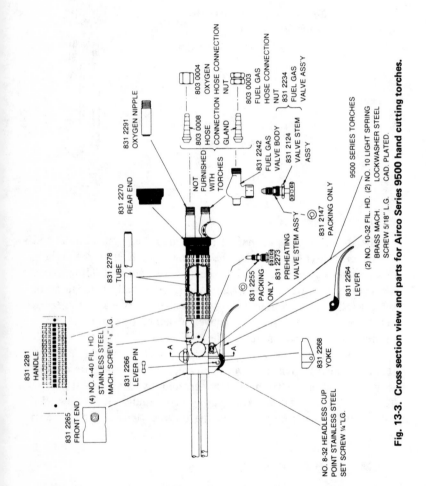

Fig. 13-3. Cross section view and parts for Airco Series 9500 hand cutting torches.

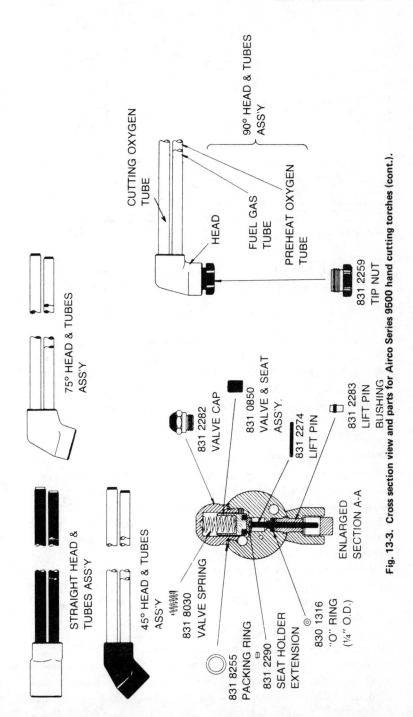

Fig. 13-3. Cross section view and parts for Airco Series 9500 hand cutting torches (cont.).

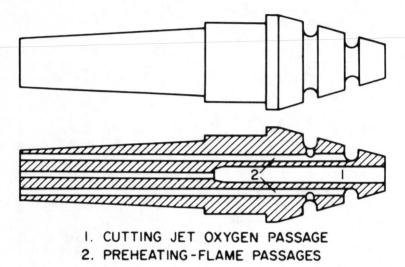

1. CUTTING JET OXYGEN PASSAGE
2. PREHEATING-FLAME PASSAGES

Fig. 13-4. Cross section view of acetylene cutting tip.

ferrous metals when they are heated above their melting point. When excess pure oxygen is added to red-hot ferrous metal, the iron will oxidize very rapidly. To say it another way, the metal will burn up. The force of the excess oxygen, in addition to burning up the metal, will also aid in blowing away the molten metal.

SUCCESS FACTORS

Needless to say, there are several factors which will influence the success of any oxyacetylene cutting process. It is important to use the proper size cutting tip for the particular cutting project at hand. All makers of oxyacetylene torches and cutting attachments offer a selection of various size cutting tips, each designed for a specific thickness of metal. Obviously, you should use the recommended tip size for any given thickness of metal.

Each cutting tip will require specific working pressures for both oxygen and acetylene. In all cases the working pressure for oxygen will be considerably more than the pressure for acetylene, as Table 13-1 illustrates.

The thickness of the metal being cut, in addition to dictating the cutting tip size, will also determine the speed at which the metal can be cut. It is important to keep in mind that the metal is heated up to the "cherry red" stage before additional oxygen is introduced. Obviously, thin metal will reach this stage much more rapidly than

Table 13-1. Pressure Chart for Sears Cutting Tips.

PLATE THICKNESS	TIP SIZE	OXYGEN PRESSURE (psi)	ACETYLENE PRESSURE (psi)
3/8-5/8″	0	30-40	5-15
5/8-1	1	35-50	5-15
1-2	2	40-55	5-15
2-3	3	45-60	5-15
3-6	4	50-100	5-15

thicker metal. It is important, then, from the standpoint of an acceptable cut for the welder to keep a sharp eye on the area being preheated and to watch for the metal to reach the proper temperature, based on the color of the metal. The preheat flames are kept operating during the entire cutting operation. The oxygen is introduced *only* when the metal has reached the proper temperature. If the excess oxygen is added before the metal has reached the proper temperature, oxidation or cutting will not take place.

As with welding, the best way to learn how to cut metal with oxyacetylene equipment is to practice often and on various thicknesses of metal until you have mastered some of the basics. It will be helpful, therefore, to run through several exercises in flame cutting and point out the important functions which must be accomplished during the cutting operation. Before we begin, however, it may be helpful to run quickly through how to set up the equipment.

SETUP

Assuming that your regulators have already been attached to both the oxygen and acetylene cylinders, and that the standard welding torch handle is also attached to hoses running from the regulators, begin by removing the welding tip attachment. In most cases this will entail unscrewing the welding tip holder from the end of the handle. At this time, the tanks should be turned off and the lines clear of both acetylene and oxygen (Fig. 13-5).

The cutting attachment is then fastened to the end of the blowpipe handle and secured. The next step is to fit the cutting attachment with the proper size cutting tip for the project at hand. For our first exercise, we will be cutting light gauge steel, less than 1/4 inch thick. Your welding outfit should have come with a booklet which recommends the proper size cutting tip for various thicknesses of metal. Choose the right size tip and fit it into the end

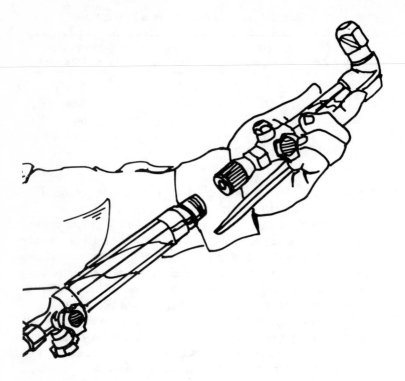

Fig. 13-5. The cutting attachment fits onto the blowpipe handle.

of the cutting attachment (Fig. 13-6). In all probability, you will be attaching a relatively small cutting tip.

Oxygen Flow

Once the proper size tip has been attached and you have double checked all connections to make certain that they are snug, open the oxygen tank valve slowly and allow some of the oxygen to pass into the regulator housing. At this time the high pressure gauge of the regulator should register how much pressure is in the oxygen tank. Open the valve slowly so that a sudden pressure will not damage the internal parts of the regulator. The diaphragm of the regulator should be wide open at this time. The low pressure or working line pressure gauge should read zero. The control valves on the torch and cutting attachment should also be closed at this time as well. The next step is to set the working pressure for oxygen. This is done by simply turning the oxygen regulator control lever

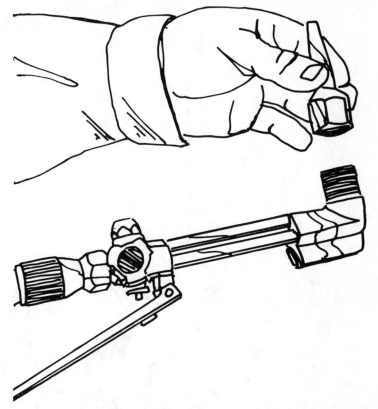

Fig. 13-6. The proper size cutting tip is fitted onto the cutting attachment.

until the low pressure gauge registers. Set the working pressure according to the recommendations for the cutting tip being used. This information can be found in the manual that came with your welding outfit (Fig. 13-7).

The next step is to give a quick check at all oxygen connections to make sure that no oxygen is leaking out into the atmosphere. If all is well, open the oxygen control valve on the welding torch handle two full turns. At this time no oxygen should flow. The reason for this is that the oxygen flow is actually controlled by the oxygen control valve on the cutting torch handle rather than the valve on the welding torch (Figs. 13-8 and 13-9).

Open the oxygen valve on the cutting torch handle a quarter turn to make sure oxygen flows. Then turn it off quickly. Next, depress the oxygen cutting lever for a second and then release it quickly. When you depress the oxygen cutting lever, a stream of oxygen should flow from the center hole in the cutting tip.

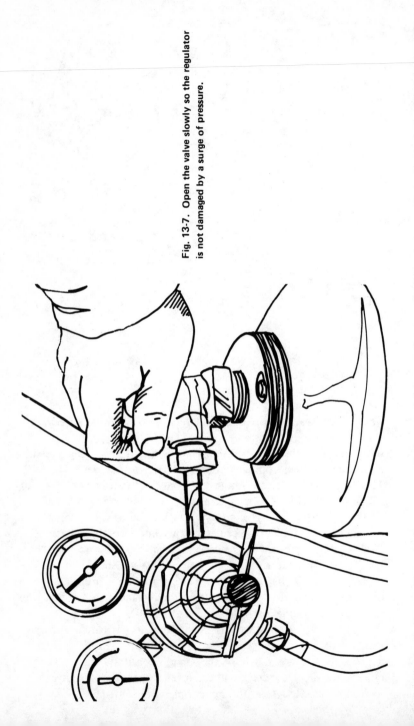

Fig. 13-7. Open the valve slowly so the regulator is not damaged by a surge of pressure.

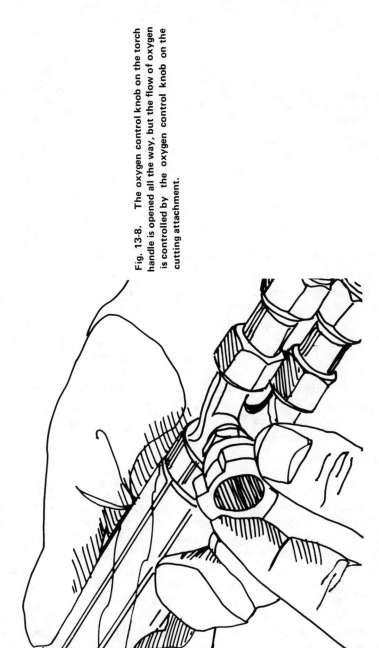

Fig. 13-8. The oxygen control knob on the torch handle is opened all the way, but the flow of oxygen is controlled by the oxygen control knob on the cutting attachment.

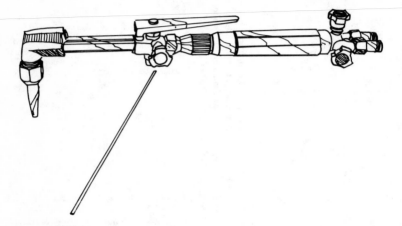

Fig. 13-9. The oxygen control knob on the cutting attachment.

After the oxygen system has been pressurized and checked for leaks and proper operation, you can turn your attention to the acetylene line. First make sure that the acetylene control knob on the torch handle is in the "off" position and the regulator control lever is also turned off (fully open). Open the acetylene tank valve slightly to let some of the fuel into the regulator housing. The high pressure gauge should register how much pressure is in the acetylene tank (Fig. 13-10).

Acetylene Flow

The next step is to turn the acetylene regulator lever until the working pressure gauge registers the proper pressure. Once again this information can be found in the booklet supplied with your outfit. After the proper working pressure has been set, open the acetylene control knob on the torch handle to make sure acetylene is flowing (Fig. 13-11). Then quickly turn it off. Give a quick check to the acetylene connections to make certain that there are no leaks.

As you can see, the only difference between setting up the oxygen and acetylene lines is that there are two control knobs for oxygen, one on the torch handle and another on the cutting attachment. There is only one for acetylene, on the welding torch handle. This may cause a bit of confusion initially, but I plan to make it clear as to which oxygen control knob is to be used when cutting.

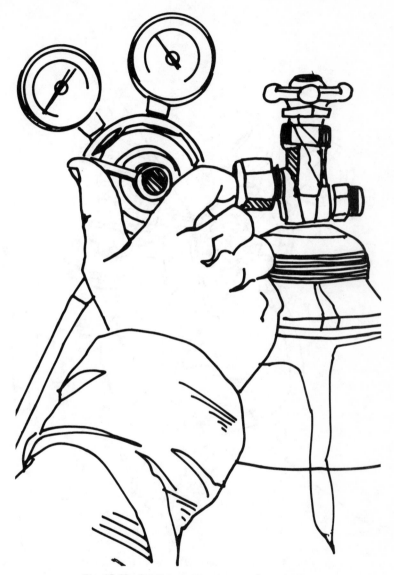

Fig. 13-10. Set the working pressure for oxygen.

The flow of acetylene is controlled by one knob on the torch handle. The oxygen knob, next to this, is opened all the way when cutting. It is important to remember that no oxygen will flow out of the torch cutting tip, however, until the oxygen flow control knob valve on the cutting attachment is opened. To summarize, the flow of acetylene is controlled by opening the acetylene knob on

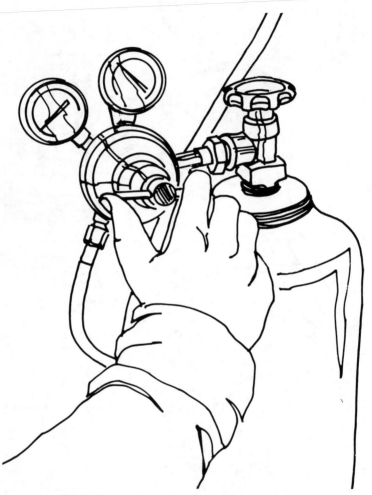

Fig. 13-11. Set the working pressure for acetylene.

the torch handle. The oxygen control knob on the torch handle is opened at all times, but the flow of oxygen is regulated by the oxygen control knob on the cutting attachment handle (Fig. 13-12).

MORE PRECUTTING PROCEDURES

The next step in learning some of the basics of oxyacetylene cutting is to lay your hands on a piece of scrap steel plate, about 1/4 inch

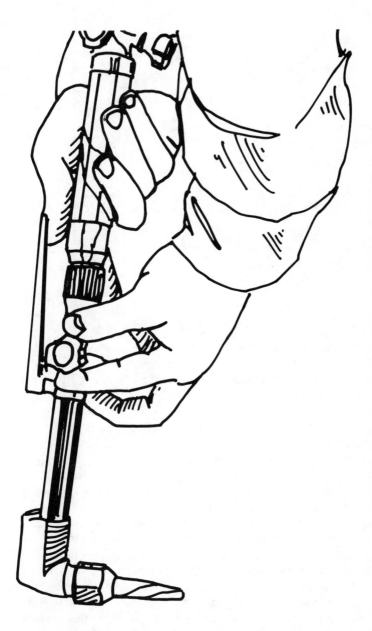

Fig. 13-12.   The flow of oxygen is controlled by this knob on the cutting attachment.

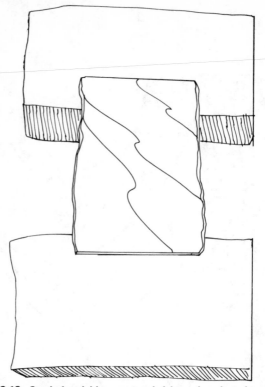

Fig. 13-13. Steel plate laid across two bricks and ready to be cut.

thick, 4 inches wide and 8 inches long. Lay this scrap lengthwise between two firebricks on your worktable (Fig. 13-13). Now, with protective goggles in place, open the acetylene control valve a quarter turn and light the torch. Then open the oxygen valve on the cutting attachment. The valve on the torch handle next to the acetylene control knob is, of course, open already. Develop a neutral flame at the preheat holes that results in a cone of about 1/8 inch long. You will find this easiest to accomplish by alternately increasing the flow of acetylene and then oxygen until the proper flame is achieved. When you are done adjusting to a neutral flame, each of the preheat holes should have this flame (Fig. 13-14).

After the neutral flame has been developed, depress the oxygen cutting lever momentarily to make certain that oxygen is, in fact, flowing through this part of the cutting attachment. You may find, when you depress the cutting lever, that the preheat flame needs to be adjusted. This is often the result of excess oxygen in the center of the preheat flames. Adjust the preheat flames if necessary. Now the torch is ready for the cutting operation.

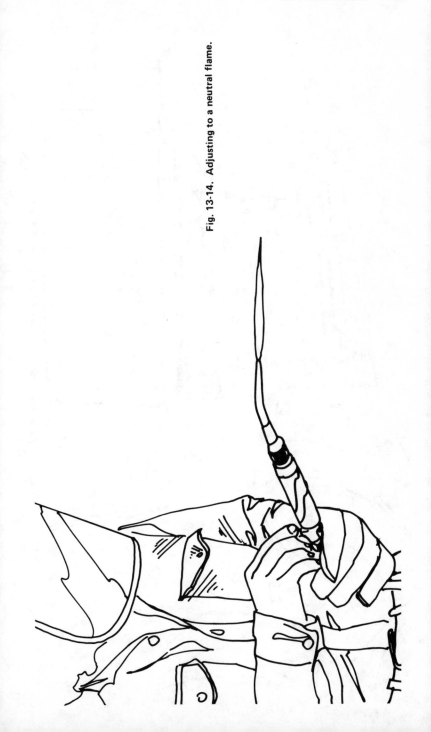

Fig. 13-14. Adjusting to a neutral flame.

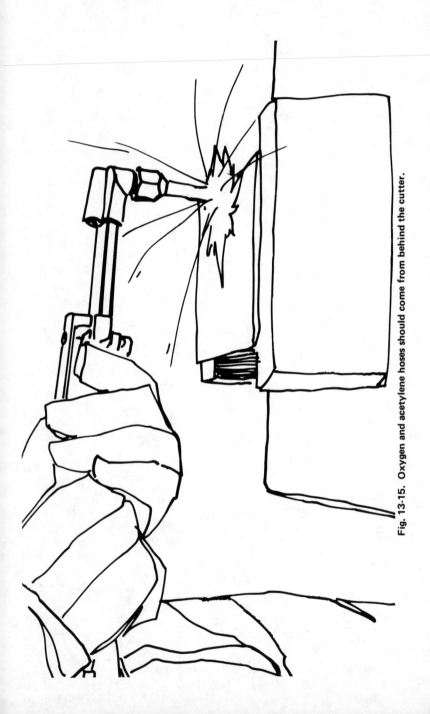

Fig. 13-15. Oxygen and acetylene hoses should come from behind the cutter.

Positioning of Hoses

Before actually beginning to cut, you should take a quick look around the work area to make sure the oxygen and acetylene hoses do not run under the work. This would be a dangerous condition especially if a spark or hot slag were to fall onto the hoses. It might cause the hose to rupture and a fire to result. For the record, it is always best for the hose to run from behind the welder. That way your body will always shield the hoses from hot sparks and molten slag (Fig. 13-15).

You should also look to be sure you have a clear path to the oxygen and acetylene tanks. If a fire were to break out, turn off the flow of oxygen and acetylene at the tanks after turning off the control valves on the torch handle and cutting attachment.

Access to Safety Equipment

Safety equipment should also be within easy reach. This includes a fire extinguisher. See Chapter 14 for solid information on fire fighting equipment. A bucket of sand and a bucket of water should also be in the welding area (Fig. 13-16).

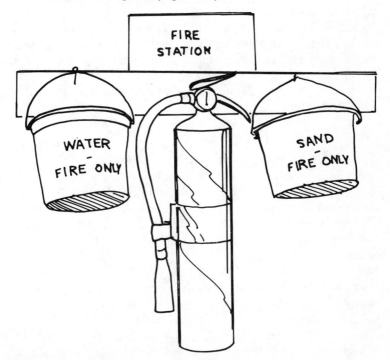

Fig. 13-16. A fire station should be within easy reach of the cutting area.

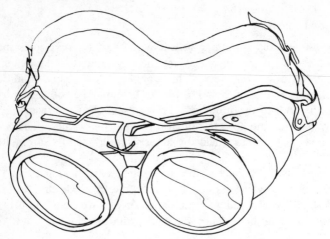

Fig. 13-17. Tinted welding goggles should always be worn when working with a torch.

As mentioned earlier in this chapter, you should have tinted goggles over your eyes. Many experts suggest that the molten metal being cut is brighter than when simply welding. Make sure your goggles are dark enough to protect your eyes (Fig. 13-17). Heavy leather gloves and other body protection should also be worn as sparks and hot slag are a common occurrence when cutting metal with an oxyacetylene cutting unit. It is important, therefore, to wear the proper clothing when cutting or welding. Chapter 14 contains extensive information about the types of clothing which are suitable to wear when working in the welding shop.

CUTTING A HOLE IN THE PLATE

To get a feel for how a cutting torch can pierce through metal, the first exercise will involve cutting a single hole in the center of the steel plate. Begin with the torch set for a neutral flame. Hold the tip of the torch about 1/8 to 1/4 inch above the center of the plate. The idea is to hold the white cones in each of the preheat flames just above the surface of the metal. In a matter of moments, you will notice the change in the color of the metal until finally the color is a "cherry red." At this time, depress the oxygen cutting lever on the cutting attachment. It is important to press the lever slowly to blow a stream of pure oxygen onto the hot metal.

As the oxygen hits the molten metal, a shower of sparks will appear at the center of the cut. If you hold the torch steady, the oxygen will remove the surface layers of the metal very quickly, exposing lower layers. These layers will almost immediately be

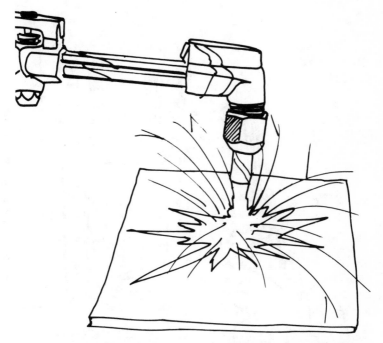

Fig. 13-18. Just before the white-hot stage, press the oxygen cutting lever.

heated by the preheat flames, and the pure oxygen will blow them away as well. In a very short period of time you will blow a hole in the center of the metal. This basically is how all oxyacetylene cutting is done. The difference between blowing a single hole through steel and cutting a straight line across a 2-foot piece of steel is simply a matter of practice and finesse (Fig. 13-18).

Some general comments about cutting are in order at this time. It is important that the cutting torch be held at a constant and steady height above the metal surface. You will find it helpful, especially in the beginning, to use two hands to hold the torch (Fig. 13-19). Assuming that you are right-handed, your right hand will hold the cutting torch at the balance point and your right thumb will be on top of the oxygen cutting lever. Your left hand will be in front of your right and will serve as a steadying point for the cutting torch. Of course, both hands should be covered with heavy gloves to protect them from hot sparks and slag.

One last reason for holding the cutting torch 1/8 to 1/4 inch off the surface is that at this height the preheat holes and oxygen

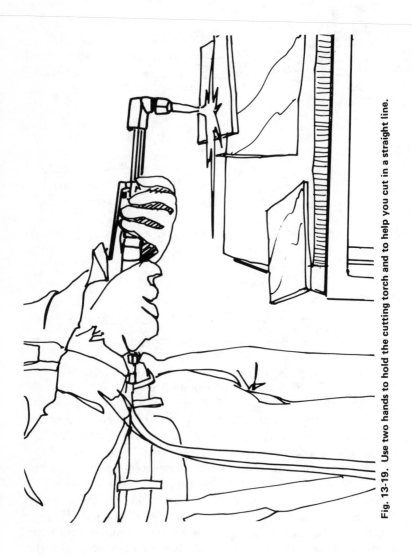

Fig. 13-19. Use two hands to hold the cutting torch and to help you cut in a straight line.

cutting hole will be protected, somewhat, from reflected heat. At this height, the holes are less likely to become clogged with flying bits of molten metal.

CUTTING A LINE ACROSS THE PLATE

The next exercise involves cutting a straight line across the steel plate. If you did not burn too many holes in the piece used for the first exercise, you can use it for this exercise as well. Before cutting, you must mark the metal so you will know where to cut. The accepted method of marking metal in the trade is to use a special *soapstone* pencil (Fig. 13-20).

Soapstone is simply a type of talc that remains visible under heat. It is used by welders for marking straight and curved cuts. Special pencils are available that can be refilled with lengths of new soapstone, and they work very much the same as a common mechanical lead pencil. One alternative to using a soapstone pencil to mark cuts on metal is to mark the cut line with a series of indentations, spaced about 1/4 inch apart, made with a center punch and a hammer. While this alternative method works quite well, it will take you quite a bit longer to mark your cuts than if you simply used a soapstone pencil (Fig. 13-21).

Mark the scrap steel plate with a line running across the piece (Fig. 13-22). Next, position the steel plate so the cut line is about midway between the two firebricks holding up the steel. Then light the torch and begin heating the metal on one edge of the cut line.

Depressing the Oxygen Lever

Once the edge of the steel plate is cherry red in color, the cutting oxygen lever is slowly depressed. This will allow a stream of pure oxygen to blast against the molten metal, resulting in sparks and the beginning of a gouge or cut.

The reason for depressing the oxygen cutting lever slowly is so the heated spot on the metal won't be cooled by the flow of oxygen. This cooling would obviously prevent cutting to take place (Fig. 13-23).

After the metal has been gouged by a fine stream of oxygen, you will notice that the hole will become deeper. It is at this time that you should press the oxygen lever once again to complete the cut. At about the same time, you should begin to move the cutting torch forward along the cut line about 1/2 inch. Since this metal is very close to that which has just been severed with the oxygen, it will be very hot. You will only have to heat it with the preheat

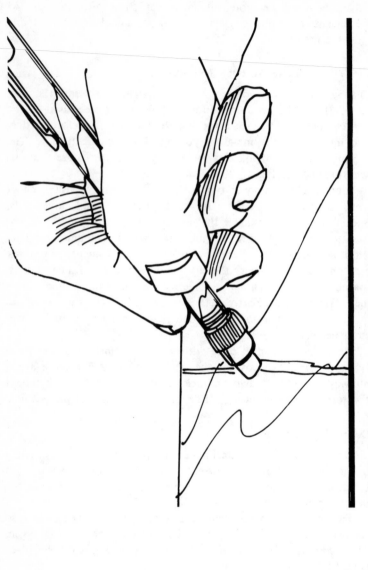

Fig. 13-20. A soapstone is the best tool for marking metal to be cut.

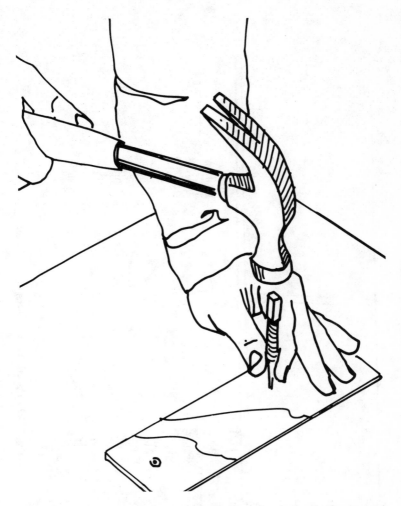

Fig. 13-21. A punch and a hammer can also be used to mark a cut line on metal, but this will take longer than using a soapstone pencil.

flames momentarily. Then spray a little of the pure oxygen onto the area. This will cut the metal. As this happens, move the torch flames forward once again. Basically, this is how all metal is cut with an oxyacetylene flame and a stream of pure oxygen.

Factors Influencing Speed

The speed at which you progress along the cut line will be determined by several factors, the thickness of the metal being the

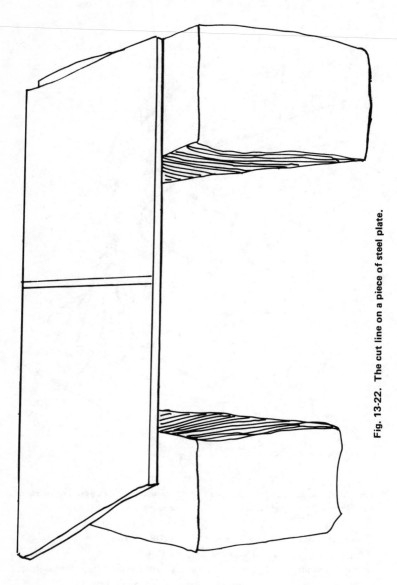

Fig. 13-22. The cut line on a piece of steel plate.

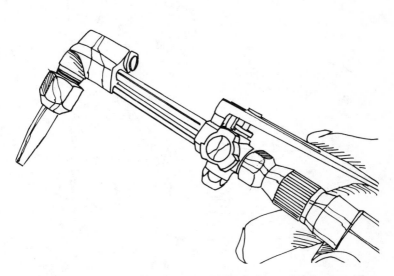

Fig. 13-23. The oxygen cutting lever should always be pressed slowly and then held when the proper cutting action is reached.

most crucial. There is an even motion when cutting metal with oxyacetylene equipment. It will take time to learn just how much heat is needed before pressing the oxygen cutting lever.

Proceed in the manner described until you have made one pass over the steel plate, along the soapstone cut line. Turn off your cutting torch and let the metal cool until it can be inspected. If you are in a hurry, you can drop the pieces into a bucket of water to cool them rapidly (Fig. 13-24). If you do this, it is a good idea to pick the hot metal up with a pair of pliers. I always keep a pair on the welding table for just such tasks.

After the metal has cooled, look closely at your first cut. There are, basically, four possibilities of how the cut will appear: perfect, too slow, too fast or too much preheat. Since each of these cuts has definite characteristics, it will no doubt be helpful to discuss them.

Perfect Cut

A perfect cut will, at first glance, appear as if the metal was cut with a hacksaw. There will be no, or very little, buildup of metal on

Fig. 13-24. Metal can be quickly cooled in a bucket of water.

either the top or bottom edge of the cut. The face of the cut will appear to have fairly straight lines running through the metal, and each will be about even with surrounding marks as far as depth. In short, a good cut made with an oxyacetylene cutting torch will look good, almost as if it were made with a machine (Fig. 13-25).

Too Fast Cut

A cut through steel plate which was made too quickly will have a buildup of slag-like metal on the bottom side of the plate. This is a result of not allowing enough time for the pure oxygen to blow

Fig. 13-25. A quality cut with oxyacetylene cutting equipment.

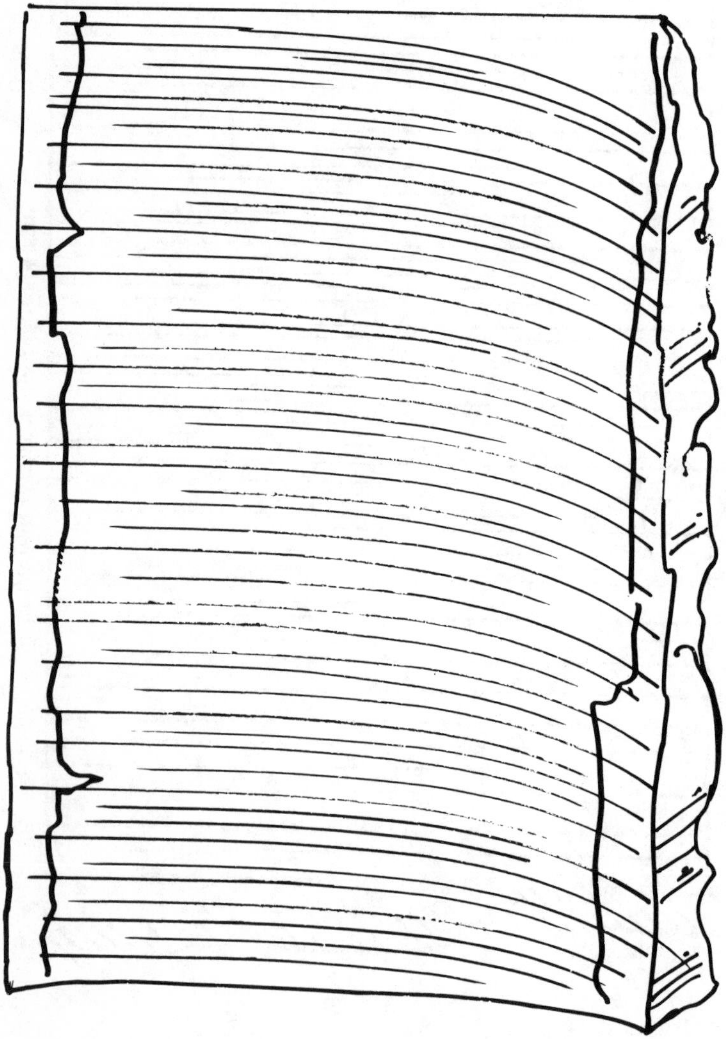

Fig. 13-26. Cutting speed too fast.

the slag out and away from the metal. The cut lines on a piece of metal that has this slag buildup will be curved, almost as if the metal were cut with a circular saw (Fig. 13-26).

Too Slow Cut

In all probability, your first cut through steel with oxyacetylene will have a buildup of slag on top, bottom and on the face of the cut. You may find that some of the cut lines are straight, while in other areas no cut lines are visible. All this is largely due to heating the metal too much. In other words, you could have moved quicker with the torch.

Another characteristic of moving the cutting torch too slowly may be burn holes in the top of the cut. If slow movement is combined with an irregular torch movement, you could end up with quite an ugly edge to the cut metal. This will appear as bubble-like mounds and very uneven cut lines on the face of the metal (Fig. 13-27).

Too Much Preheat Cut

When more heat than is required is applied to the cut line before depressing the oxygen cutting lever, the top of the metal will indent. This happens because more of the metal along the cut line is ready for the chemical reaction of the oxygen. In extreme cases of too much preheat, the top edge of the cut will appear to flow down the face of the cut. In effect, this is exactly what has happened (Fig. 13-28).

PROFESSIONAL TIPS

As you can see from the description of the various cuts, the speed, height and intensity of the cutting torch preheat flames are points over which you should have the most possible control. It will be helpful to mention a few professional tips to help you make good clean cuts in metal with an oxyacetylene cutting torch.

Observe the Metal

One of the most important things that can be learned by beginners is to keep a very close eye on the metal. As it heats up to about the right temperature, you should be ready to give it some oxygen. The novice is really at a disadvantage here even if he or she has a keen eye. Until you have watched metal heat up with an oxyacetylene flame a few times, you will not really know what to look for. You must be aware that the metal will move up in temperature

Fig. 13-27. Cutting speed to slow.

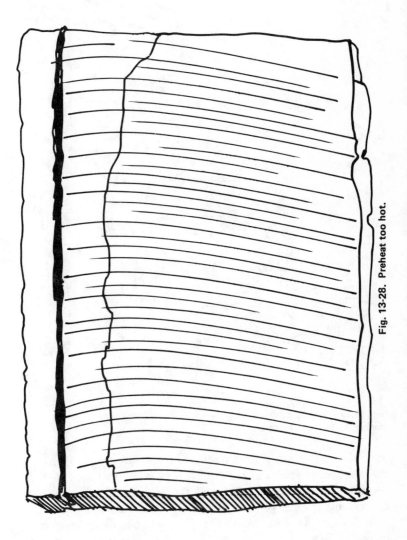

Fig. 13-28. Preheat too hot.

Fig. 13-29. You must depress the oxygen cutting lever just before the metal reaches the white-hot stage.

quite rapidly. It will change from gray to red quickly and then pass through several shades of red until it is practically white. If heat is constantly applied, the metal will turn white and become truly molten, just as when you were puddling metal during welding.

When the metal has become molten, without first adding pure oxygen, it will not cut well. When the oxygen is introduced at this point, the metal will deteriorate very rapidly. The time to depress the oxygen cutting lever is before the white-hot stage of the metal (Fig. 13-29). Only then can you expect to cut the metal cleanly.

Adding Oxygen

The manner in which you depress the oxygen cutting lever will have a very real effect on how the metal cuts, and even on how the metal will look after being cut. It is a common fault of beginners to add too much oxygen when cutting. This is understandable but not acceptable.

The proper way to add oxygen to cherry red metal is to very slowly press on the oxygen cutting lever until the metal is flowing away at an even stream. This is actually a very soft, subtle touch that can only be developed in time. In the beginning you will more than likely press the oxygen cutting lever all the way down as soon as you see that the metal has reached the right color. In time, however, you will learn that only a bit of oxygen is needed. You will develop the soft touch necessary to make a good, clean cut.

Proper Torch Movement

Keeping the cutting torch moving along a straight line, while at the same time keeping the preheat flames at the same height above the metal, is another requirement for a good cut. As I mentioned earlier, you should use your left hand as a steadying point for the torch while controlling the oxygen cut lever with your right hand and thumb. This is an awkward position when you have to move the torch to cut along a straight line. There will be a natural tendency to put a curve in the cut. One way to overcome this is to move your left forearm and right hand along the line. The task will be easier to accomplish if you also move your whole body at the same time. A good work table will help you to move in front of the work while cutting (Fig. 13-30).

One way to eliminate much of the body movement necessary to make a straight cut is to attach a temporary guide for the cutting torch. Then you can simply rest the barrel of the cutting torch on the guide. Move the torch along the cut line at both the right height and along a straight path.

A guide for any oxyacetylene cutting torch can be fashioned quite simply from pieces of scrap steel you should have around the shop, like a piece of angle iron (Fig. 13-31). The only requirements of the metal used for the guide are that it must be straight, at least as long as the width of the cut being made and, probably most important of all, the right height for the torch being used.

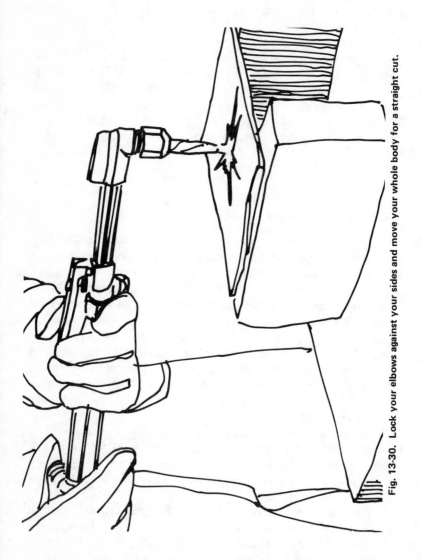

Fig. 13-30. Lock your elbows against your sides and move your whole body for a straight cut.

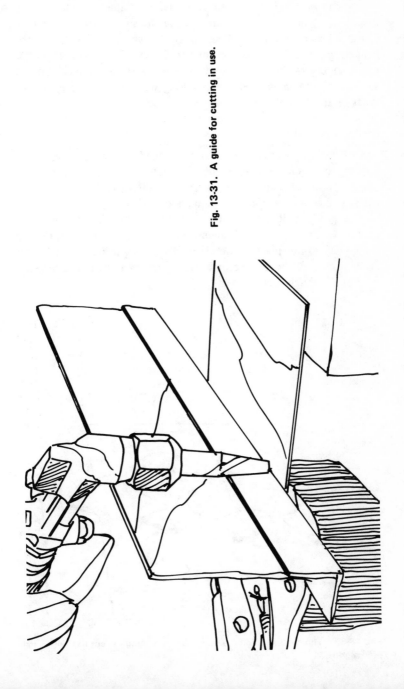

Fig. 13-31. A guide for cutting in use.

Preheat Flame Height

As you are probably aware, the preheat flames should be at a height of about 1/8 to 1/4 inch above the surface of the cut line. If the white cones are too high above the work, the metal will not heat properly or will take too long to do so. On the other hand, if the heat cones are too close to the work, you will heat up the surface of the metal too much, along with possibly clogging the holes on the face of the cutting tip.

Setting Up a Guide

Obviously some experimentation is called for on your part when setting up a guide for your cutting torch. Once you have found angle iron which is a suitable height for the torch and tip you are using, you should be able to obtain good results.

The guide, if it is angle iron, can be quickly fastened in place parallel to the cut line with clamps (Fig. 13-32). When positioning the guide, do not place it too close to the area to be cut or too far away for it will be ineffectual. Most experts agree that a distance of about 3/4 inch is about best.

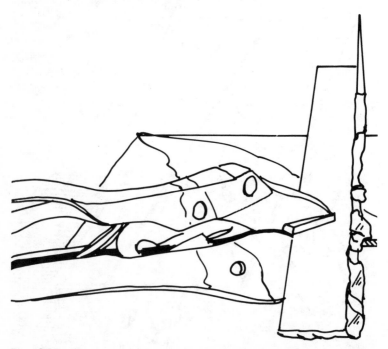

Fig. 13-32. The cutting guide is simply clamped along the cut line with a pair of vise-grips.

A guide for an oxyacetylene cutting torch will enable you to make straight cuts on metal. But you still must keep a close eye on the metal so you will know when to depress the cutting lever. You will also have to proceed along the cut line at a speed which is adequate for the thickness of metal being cut.

You should practice cutting straight lines in 1/4 inch thick steel plate until you can make perfect cuts. In the beginning, use some type of guide for the cutting torch. This will enable you to develop a technique with a little more ease. After you have mastered this, try cutting without the aid of a guide for the torch (freehand).

THIN METAL CUTTING

Thin metal can also be cut with an oxyacetylene cutting torch. Thin metal is any type which is less than 1/4 inch thick and is most commonly referred to as sheet metal. For cutting sheet metal, a different approach is needed as there is always a tendency to burn holes in the metal. Two professional pointers will be of use to you when cutting thin metal.

Hold the torch at an angle to the cut line on the surface of the metal. A 45 degree angle, or less, for very thin metal will enable you to make the cut without leaving large holes in the cut line (Fig. 13-33). Another point to keep in mind is to work quickly. The more time you spend heating the metal, the more the chances of burning holes rather than cutting a straight line.

When cutting sheet metal, you should use the smallest possible cutting tip. Some experts can make cuts in metal of this type with

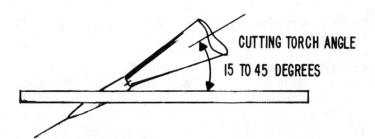

CUTTING TORCH ANGLE
15 TO 45 DEGREES

Fig. 13-33. Cutting thin sheet metal with an oxyacetylene cutting torch. The torch angle is slight to moderate.

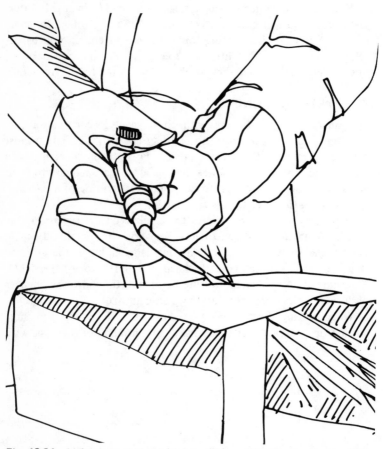

Fig. 13-34. Light sheet steel can be cut with a welding torch. A very slight angle is used for the torch.

a large welding tip and a slightly oxidizing flame (Fig. 13-34). In time, you may want to experiment with cutting with a welding tip also. For now, use a small cutting tip.

BEVEL CUTTING

As you may recall, steel which is thicker than 1/4 inch must have beveled edges or it will be difficult to weld with the oxyacetylene

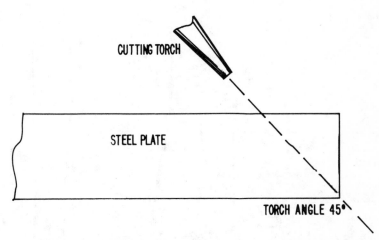

Fig. 13-35. Cutting a bevel edge on steel plate with an oxyacetylene cutting torch.

process. In Chapter 10 directions were given for grinding the edges of metal to develop a bevel. You may have already tried this and discovered that it takes quite a bit of grinding to bevel the edges of plate steel. Another way to bevel edges of steel plate is, of course, to cut a bevel with a cutting torch (Fig. 13-35). While this type of cutting is not difficult, it does require a slightly different approach than for straight cutting.

The major difference between straight cutting of steel plate and bevel cutting of the same material is the angle at which the cutting torch is held. To cut a bevel, the torch tip must be held at a 45 degree angle to the cut line rather than perpendicular (Fig. 13-36). In most cases, some type of guide for the cutting torch will enable you to cut at both a constant bevel and along a straight line.

Another helpful tip when cutting a bevel on steel plate is to use a slightly oxidizing flame. To refresh your memory, first develop a neutral flame for the preheat holes and then add a bit more oxygen.

It is important, when making a bevel cut, to keep the torch moving at a consistent speed. An unsteady cutting speed will often result in an irregular cut and possibly a stop in the cutting action.

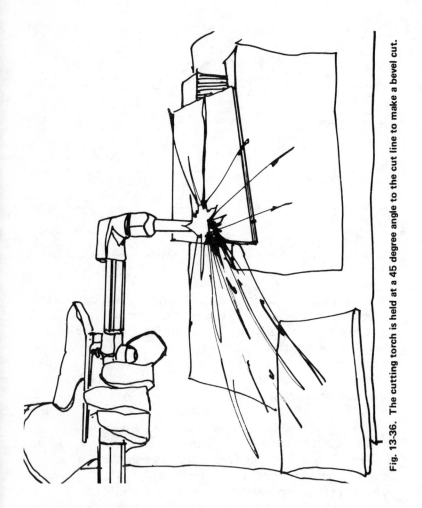

Fig. 13-36. The cutting torch is held at a 45 degree angle to the cut line to make a bevel cut.

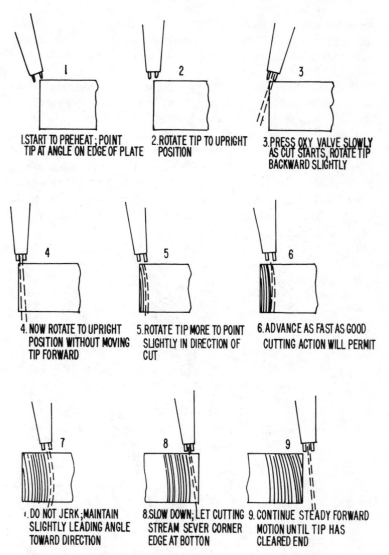

1. START TO PREHEAT; POINT TIP AT ANGLE ON EDGE OF PLATE

2. ROTATE TIP TO UPRIGHT POSITION

3. PRESS OXY VALVE SLOWLY AS CUT STARTS, ROTATE TIP BACKWARD SLIGHTLY

4. NOW ROTATE TO UPRIGHT POSITION WITHOUT MOVING TIP FORWARD

5. ROTATE TIP MORE TO POINT SLIGHTLY IN DIRECTION OF CUT

6. ADVANCE AS FAST AS GOOD CUTTING ACTION WILL PERMIT

7. DO NOT JERK; MAINTAIN SLIGHTLY LEADING ANGLE TOWARD DIRECTION

8. SLOW DOWN; LET CUTTING STREAM SEVER CORNER EDGE AT BOTTOM

9. CONTINUE STEADY FORWARD MOTION UNTIL TIP HAS CLEARED END

Fig. 13-37. Recommended procedure for efficient flame cutting of steel plate.

THICK STEEL CUTTING

Cutting thick steel can also be done with an oxyacetylene cutting torch. Generally speaking, the thicker the metal, the longer it will take to make a cut. The cutting operation is basically the same as those previously discussed with a few modifications. One of the first things to keep in mind is that the cutting torch tip must be held closer to the work surface of the metal. In most cases this will mean 1/16 to 1/8 inch above the cut line. It is also important that the torch tip be held at a right angle to the cut at all times. This will ensure that the heat is being applied evenly as the cut progresses (Fig. 13-37).

Probably the most difficult part about cutting thick metal with oxyacetylene is starting the cut. There are several different ways of starting which are used by professional welders.

One method of starting, assuming that you are right-handed, is to begin at the edge of the metal and move from left to right. This will give you a clear view of the cut and let you look into it to make sure the slag is being blown out by the oxygen. When moving from left to right, you may have some difficulty in following the cut line. You can usually overcome this by using some type of guide for the cutting tip.

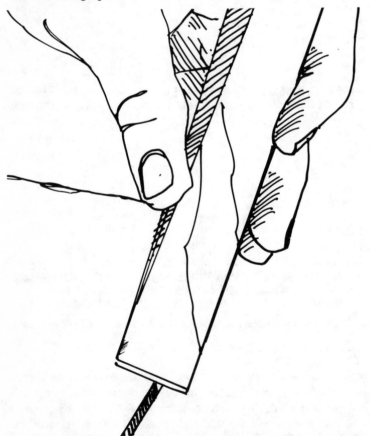

Fig. 13-38. Some professionals like to chip the edge of thick steel plate before starting the cut.

Another method used to start a cut in thick metal is to begin at the corner of the metal by slanting the torch in a direction opposite the direction of travel. As the corner heats up to cherry red and is cut, the welder then moves the torch to a vertical position until the steel has been cut. Then proceed along the cut line.

Some professional welders like to chip the edge of the cut line with a cold chisel and hammer before starting the cutting operation (Fig. 13-38). The sharp edges of the chip will preheat and oxidize more rapidly after doing this.

One last method of starting a cut involves using an iron filler rod. The rod is placed along the cut line before starting (Fig. 13-39). Then, as the preheat flames are applied, the filler rod will reach the cherry red stage very quickly. As soon as this happens, the cutting oxygen lever is depressed causing the filler rod to oxidize. A chain reaction then takes place, and the parent metal begins to oxidize at about the same time.

Whichever means you use for starting a cut, and you should try them all, it is important to move the cutting torch at a steady rate once you begin the cut. After the edge of the metal has been cut, the oxygen cutting lever should be kept constantly open by pressing it lightly. As the metal heats up along the cut line, more cutting oxygen is added. At the same time move the torch slowly and steadily along the cut line.

PIPE CUTTING

One last use of the oxyacetylene cutting torch that we will discuss is cutting pipe. After all, how can you possibly expect to go to Alaska to work on the pipeline as a welder if you have no idea of how pipe is cut? Think of all the salmon fishing you would miss!

Pipe Marking

Before metal pipe can be cut, it must be marked with a soapstone pencil and a special marking guide. Any welding supply house will sell special pipe marking paper, but you can usually find something around your shop that will do the job just as well. What you need is a strip of stiff paper, such as matboard, about 1 to 2 feet long and about 6 inches wide (Fig. 13-40). This paper is then wrapped around the pipe until it overlaps. Make sure the edges overlap evenly and then mark the pipe for cutting.

After the pipe has been clearly marked for cutting, place it on your work table. Prevent the pipe from rolling off by placing it

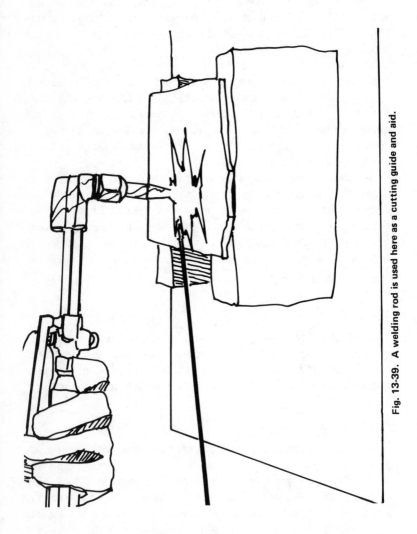

Fig. 13-39. A welding rod is used here as a cutting guide and aid.

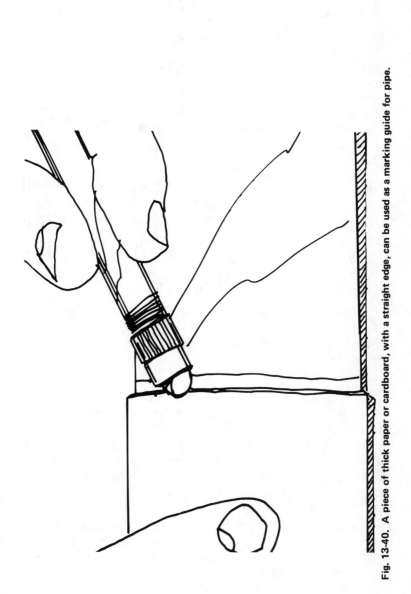

Fig. 13-40. A piece of thick paper or cardboard, with a straight edge, can be used as a marking guide for pipe.

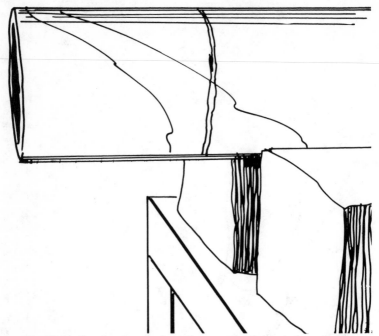

Fig. 13-41. Steady a length of pipe between two bricks before cutting.

between firebricks (Fig. 13-41). The area to be cut should extend off the table.

Torch Angle

The size of the pipe will determine for you the angle at which you hold the cutting torch. For pipe less than 4 inches in diameter, you should hold the torch so that the preheat flames are almost tangent to the circumference of the pipe (Fig. 13-42). For pipe thicker than 4 inches, you can usually hold the torch perpendicular to the pipe without danger of burning through the other side of the pipe (Fig. 13-43).

The thickness of the walls of the pipe, rather than the diameter of the pipe, will determine the size cutting tip to use in the torch. Consult the guidebook that came with your torch for the proper tip size recommendations.

Two-Man Operation

Begin cutting anywhere along the cut line on the pipe. As the metal reaches the cherry red stage, press the oxygen cutting lever slowly until you get a good cutting action going. As the actual

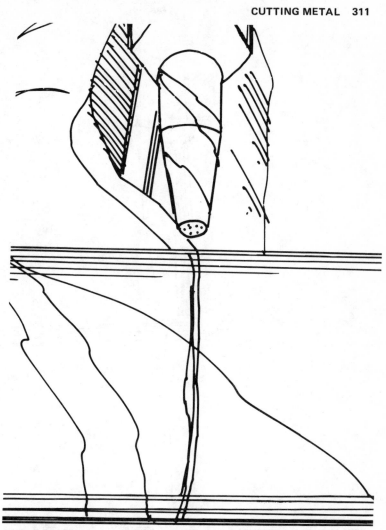

Fig. 13-42. Cutting small pipe less than 4 inches in diameter requires only a slight torch angle.

cutting begins, you must either move the torch along the cut line or have someone else turn the pipe for you. In most instances, cutting pipe will be easier if it is a two-man operation. One person does the cutting and another turns the pipe. Needless to say, both welder and turner should wear suitable clothing and eye protection.

Beveling Edges

Beveled edges are also possible when cutting pipe. You will find the work a bit easier to accomplish if you point the torch

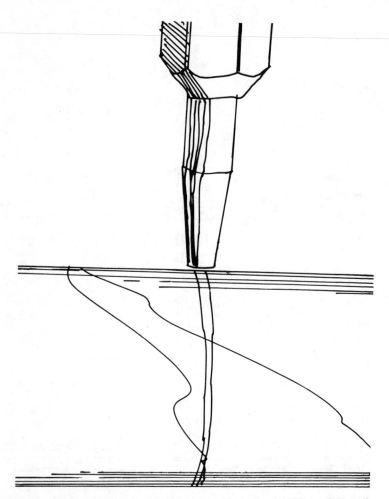

Fig. 13-43. Pipe larger than 4 inches is cut with the torch held at a right angle to the cut line.

towards the short end of the pipe as you proceed with the cutting (Fig. 13-44).

If you find that you must work alone when cutting pipe, and the pipe is manageable by one person, you can do the cutting on the ground and use your foot as a means of holding the pipe (Fig. 13-45). Needless to say, you should exercise extreme caution when doing this type of cutting. It is almost as much a balancing act as a cutting operation. Heavy work boots are also required.

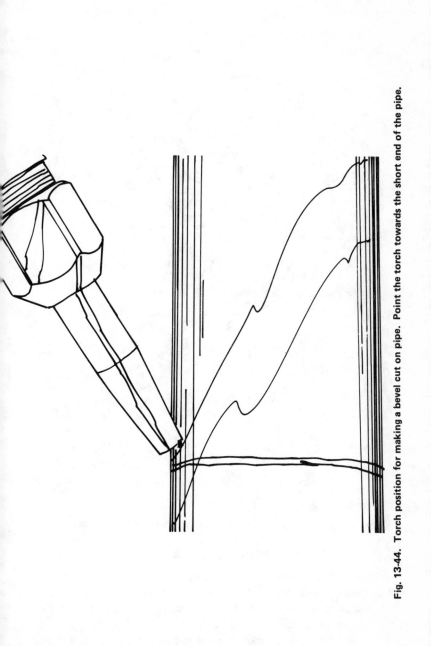

Fig. 13-44. Torch position for making a bevel cut on pipe. Point the torch towards the short end of the pipe.

Fig. 13-45. If you are very careful, you can cut pipe in this fashion.

PIERCING A HOLE THROUGH THE PLATE

The first exercise in this chapter involved burning a hole in the center of a piece of steel plate. At some point in time, you may find that you require a hole in a piece of metal. It will be to your distinct advantage to know how to approach the problem. The process is really quite simple and can be mastered in a short period of time. Begin by choosing a cutting tip that is one size larger than that recommended for standard cutting on the particular thickness of

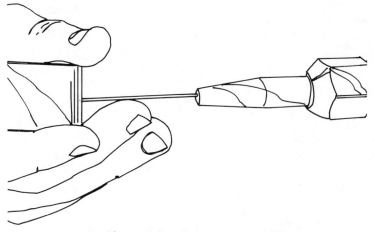

Fig. 13-46. A clean cutting tip works best.

metal you are working on. Next, hold the lighted tip of the cutting torch 1/16 to 1/8 inch above the surface of the metal. It is important that the tip be held at a perpendicular angle.

Heat the spot until it is cherry red and then very slowly depress the oxygen cutting lever. At the same time, the nozzle of the cutting torch should be raised slightly off the surface, say 1/2 inch. This will reduce the chances of slag being blown up into the nozzle holes and clogging them. As the cutting takes place, you may find it necessary to lower the tip slightly and then raise it once again. In a matter of moments, a hole should be blown through the plate. When this happens, lift the torch up and away from the work.

Piercing a hole in steel plate always runs the risk of clogging the holes in the cutting tip. You should therefore keep a close eye on the flames. If they appear erratic, chances are very good that bits of slag are building up on and around both the preheat and oxygen cutting holes. If this happens, you should stop work and clean the tip with a special tip cleaner (Fig. 13-46).

Oxyacetylene cutting is a dramatic experience because of the volume of sparks produced and the speed at which metal can be severed. It will take several hours of practice before you can cut well, but it is a welding skill worth knowing. There is something almost magical about working with molten metal, even though there are inherent dangers.

In the next chapter we will discuss how to safely accomplish many welding and cutting tasks. It will be to your advantage to make safe welding and cutting the only way to know how to work.

Chapter 14
Safety

From the time we enter into this world until the last beat of our heart, there will always be some type of danger. Depending on how and where we live our lives, the way we get from one place to another and how we accomplish various tasks, we always run the risk of injury or worse. That is the natural scheme of things.

The home welder, in addition to all the normal risks to health, lives just a bit closer to danger because of the nature of welding equipment. Any welding or cutting operation should never be considered completely safe because to do so would make working with extreme heat second nature, which it surely is not. If you forget for a moment that you are working with a flame that can be as hot as 6,000 degrees Fahrenheit, or that you are around pressurized containers containing explosive gases, you are flirting with mishap or disaster to yourself and others. While it is impossible to eliminate all of the dangers of working with a torch and hot metal, it is possible to reduce the inherent risks by developing safe working habits and religiously following accepted and recommended procedures for handling the equipment.

The purpose of this chapter, then, is to make you aware of the dangers of working with welding equipment, specifically the materials and tools for oxyacetylene welding. With this information in the back of your mind, you should be able to perform all welding tasks with the distinct advantage of knowing how things in general can be safely done. By knowing and following these guidelines, you will be able to enjoy metal joining and protect yourself from harm.

CLOTHING

Since you cannot control the direction that sparks fly when welding or cutting, you must protect your skin from possible damage. The best way to do this is to wear specially made clothing that will not support combustion. This type of clothing is available, but it is also

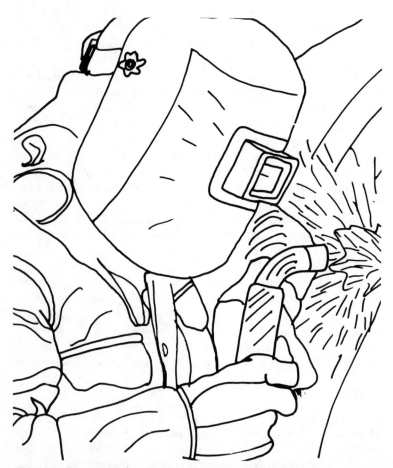

Fig. 14-1. Since a welder is always close to the work, it is important that protective clothing be worn.

quite expensive. Even so, if you plan to do a good deal of welding, it may be to your advantage to invest in fireproof clothing. It is often a requirement for industrial type welding (Fig. 14-1).

Cowhide Garments

As you may have guessed, there are alternatives for the home welder in the clothing department. Probably the best choice are garments made from thick cowhide (Fig. 14-2). Currently available are jackets with long sleeves and a high collar which effectively

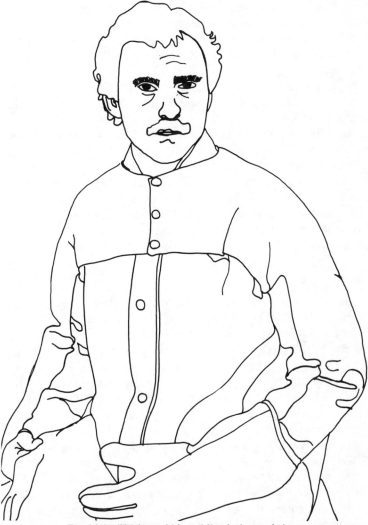

Fig. 14-2. Thick cowhide welding jacket and gloves.

protects the top half of your body from burns. Leather pants are also sold. While they amount to a sizable investment, about $80 to $100 per suit, this combination is a very good way of protecting your body from the hazards of burns from sparks and flying hot slag.

Cotton and Wood Apparel

Cotton and wool clothing are undoubtedly the most popular types with both casual and professional welders alike (Fig. 14-3). The ever trusty Levi trousers are heavy enough to shed sparks and will not burst into flames if subjected to open flame. They will smolder, however, but this does not seem to be a problem since the area can be quickly extinguished with a brush of a gloved hand or a few drops of water.

It is important that all clothing worn during welding be free of any trace of oil or grease. These materials burn violently in the presence of pure oxygen.

Wool, in the form of heavy long-sleeved shirts, is quite popular with welders as it is nonflammable, lightweight and has a relatively long life. Both cotton and wool clothing can be washed frequently without appreciable loss of the material's life. Regular and periodic washing of welding clothing should be done, not as much for sanitary reasons but for safety reasons. The material itself will not burn, but some forms of dirt or grime will and should be removed as often as necessary.

Flameproof Clothing

You can flameproof your welding clothing quite simply by soaking in a special fire retardant solution. First mix up a batch of 12 ounces sodium stannate to 1 gallon of warm water. Then dip the clothing in this mixture and let it soak for about 15 minutes. Next, wring out the material and then dip into a solution of 4 ounces of ammonium sulfate to 1 gallon of water. After about 15 minutes, remove from the solution, wring thoroughly and let dry. Clothing treated in this manner will be flameproof and quite safe to wear when welding or cutting.

Special care is required for flameproof clothing. Only wear the clothing for welding and not for working on the car or truck. When not in use, hang the clothing in an area where it will remain clean. If the clothing ever becomes dirty, as it surely will, do not wash it as this will remove the fireproof impregnation. Instead, have the clothing dry cleaned. It should last for about six cleanings.

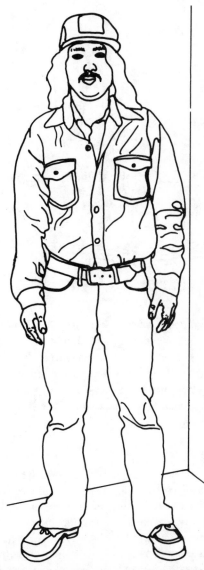

Fig. 14-3. A hat, wool shirt and heavy pants are suitable for most types of oxyacetylene welding.

Fig. 14-4. Cuffs on trousers can be dangerous. Never weld or cut with pant legs rolled up.

Whenever you are wearing clothing that is not specifically designed and made for welding, you should know that cuffs on pant legs and pockets on both trousers and shirt are potentially dangerous. These parts of clothing have a tendency to catch and hold hot sparks. It is therefore a very good idea to roll down cuffs and remove pockets, especially on shirts or, at the very least, button them closed (Fig. 14-4).

Head Covering

Some type of head covering should be worn by the welder as well as those types of clothing already mentioned. A baseball cap or similar type will not only keep your hair clean, but serve as a protection against hot sparks as well. Since most welding and cutting is done at less than an arm's length away, there is always the possibility of being struck by sparks. While there is little the welder can do to prevent this, he can and should take precautions to protect himself from the hazards of red-hot sparks.

Fig. 14-5. Tinted goggles and a baseball cap are a good combination for head protection during welding.

A baseball type cap is most effective if worn with the bill on the back of the neck rather than in the conventional manner. This will offer added protection against a hot spark falling down your neck (Fig. 14-5).

Boots and Shoes

When thinking about clothing for the welder, it is also important to consider what is worn on the feet. Heavy duty leather boots are the best choice with those without laces at the top of the list. There are any number of quality shoes and boots designed for use during the welding process. Some styles have the added feature of metal toe inserts which prevent damage to the foot from falling heavy objects.

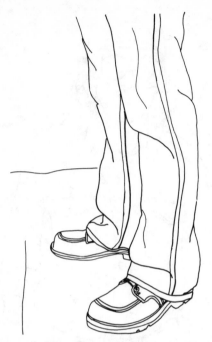

Fig. 14-6. Heavy work shoes or boots should be worn when cutting or welding.

Lightweight shoes or, heaven forbid, canvas shoes should never be worn when welding for obvious reasons. As mentioned, boots are the best form of protection, but heavy work shoes can also be worn providing that the pant legs cover the tops of the shoes well. Since you will, in most cases, be devoting your energy to the area being welded, you may not notice a hot spark burning a hole in the top of your shoe. This is another very good reason for wearing heavy duty footwear (Fig. 14-6).

EYE PROTECTION

Surely one of the cardinal rules of all welding is, "Never look at a welding flame or arc unless you have suitable protection covering your eyes." Eye protection is not only necessary to prevent damage to the eye from the injurious rays created during welding or cutting, but also to protect against sparks and hot slag.

Goggles

There is a wide selection of good eye protection gear available to anyone planning to weld. In fact, most welding outfits will come with a pair of special goggles in the package. Goggles are available in

Fig. 14-7. Light duty welding goggles.

both high impact plastic and safety glass. **Given** the choice, most professionals will choose the safety glass type. These goggles do not scratch and tend to offer an undistorted view of the work. Any quality pair of safety goggles has lenses which can be removed and replaced if the need should ever arise (Figs. 14-7 and 14-8).

Face Shields

Also available are face shields in both clear and tinted styles. Face shields have a few advantages over goggles that are worth mentioning. The field of view is greater and will enable the welder to move about more naturally. While wearing goggles, the welder may feet quite comfortable while working on a project in a fixed location. Movement with goggles on will give the feeling of tunnel vision.

A face shield offers more protection than a pair of goggles simply because the entire face and part of the head is covered (Fig. 14-9). Every face shield I have ever seen could be easily lifted up and away from the face, offering the chance for quick, normal vision if the need should arise. A pair of goggles, on the other hand, may not be quite so easy to remove quickly, especially when a hat or helmet is worn along with them.

One of the disadvantages of a face shield is that they are all made from a plastic material, very much the same as a face shield for a motorcycle helmet. Plastic will scratch quite easily. Therefore, the welder's field of vision may become distorted somewhat.

Filtering Lens

It is a fallacy to think that any pair of tinted glasses or goggles can be used for all types of welding or cutting. A special lens is needed to filter out harmful rays when welding. This filter must be of a type that is suitable for the particular welding or cutting task at

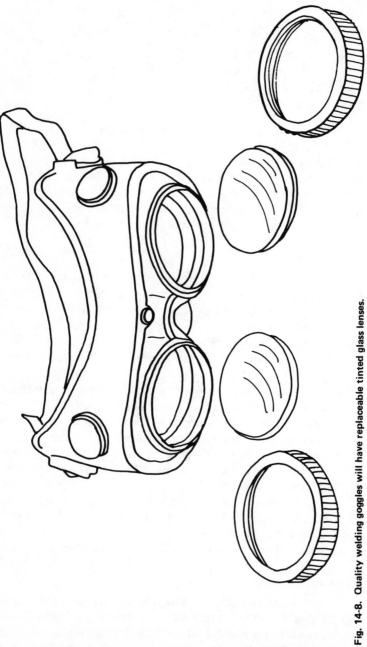

Fig. 14-8. Quality welding goggles will have replaceable tinted glass lenses.

Fig. 14-9. A tinted face shield designed for oxyacetylene welding and cutting.

hand. A filter's job is to absorb ultraviolet and infrared rays as well as most of the visible light rays. It is important, then, that you use only goggles or other eye protection that has been specifically designed and approved for welding or cutting.

If you wear corrective lenses, speak to your optometrist about buying a special pair of welding goggles tailored to your particular vision requirements. Also available are tinted goggles that fit over conventional eyeglasses.

Clear Gear

In addition to a pair of tinted welding goggles, the home welder should have some type of clear eye protection. In most cases, a clear face shield will be more than adequate. Use clear eye protection gear when grinding metal to prevent eye damage (Fig. 14-10). Never rely solely on the shields that are common on most quality bench grinders as these will do little, if anything, to protect your eyes from flying sparks and metal particles.

Fig. 14-10. A clear face shield should be worn when grinding.

It is a sound principle to keep your clear eye protection gear right next to your bench grinder or other abrasive machinery. Then get into the habit of putting on the eye guard before turning on the machine. This small act will go a long way in protecting your eyes from damage.

RESPIRATORS

Occasionally a welder may come in contact with fumes (both non-toxic and toxic) as well as normal air that is heavy in particulate matter. Probably the most common time for this contact is when grinding metal on a bench grinder, wire brushing metal or when working with a disc sander. While in most cases this "heavy air" will do little harm to the welder, some measures should be taken to protect the respiratory system.

There are a number of different types of respirators available. The range goes from a bandanna secured over the nose and mouth to self-contained breathing masks that would be quite effective during a poisonous gas attack. A little common sense should come into play here. For example, if you are grinding metal that has a lot of surface rust and you can see that you will be working on the piece for a period of time, you should wear an adequate covering over your mouth and nose.

One of the more popular means of protecting the respiratory system is an inexpensive cloth mask (Fig. 14-11). You can pick one of these masks up at most welding supply houses. They are also commonly sold in paint stores and lumber yards. Since their cost is minimal at the present time, about three for a dollar, you should pick up a supply and keep them handy around your welding shop. Get into the habit of using a respirator whenever you must work in an area that has a heavy concentration of particulate matter in the atmosphere. This will reduce the chances of serious respiratory damage as a result of breathing this type of air.

GLOVES

Another cardinal rule of welding is, "Never pick up a welding blowpipe or cutting torch without a pair of heavy duty gloves on your hands." By following this simple rule, you will dramatically reduce the chances of burning your hands.

Since at least one of your hands will always be within 1 foot of a flame with a temperature of up to 6,000 degrees Fahrenheit, you should always protect your hands from sparks and reflected heat.

Fig. 14-11. A respirator is a good precaution when sanding or wire brushing rusted metal.

The best way to do this is to wear a heavy duty glove on each hand (Fig. 14-12).

There are easily hundreds of different types and styles of gloves designed for welding. Add to this amount a vast selection of general purpose heavy duty work gloves, and you will have little problem in finding a pair of gloves that will effectively protect you while welding or cutting. It is important to keep in mind, however, that not just any pair of gloves is suitable for working around hot metal.

The gloves you wear for welding can be a bit lighter than those you use for cutting. But, in all cases, the gloves should be heavy and practical. There is a cutoff point. A pair of gloves that are more than adequate for working with a torch may also be too heavy for effectively controlling the flow of oxygen or acetylene. In other words, the gloves may not allow you to turn the control knobs on the blowpipe handle. Consider this carefully when shop-

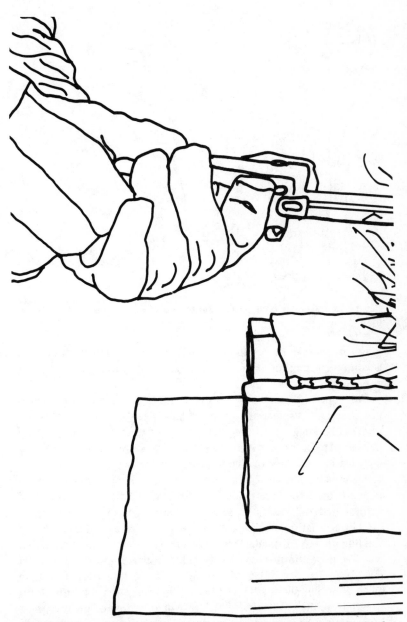

Fig. 14-12. A welder's hands are always close to the heat, so a good pair of gloves is very important.

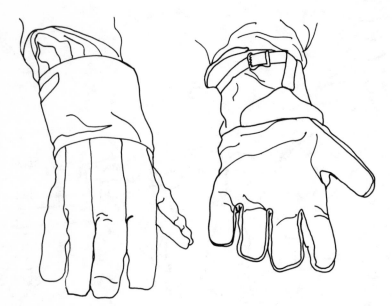

Fig. 14-13. Gauntlet type gloves offer the most protection against sparks and hot slag.

ping for a good pair of gloves. Purchase a pair that will offer both adequate protection from the heat and will allow you to adjust equipment controls with your fingers.

All experts agree that a pair of welding gloves should be flame-proof and of the gauntlet type (Fig. 14-13). You will have no problem finding these longer type gloves at any welding supply house. At the present time, welding gloves are made from heavy duty leather, asbestos or a combination of the two.

Personally, I have found that a pair of welding gloves are easier to work in if they are made from leather. Leather, because it is a natural material, tends to facilitate hand and finger movement a bit better than man-made fibers. Many styles and thicknesses of welding gloves are made from leather. Generally speaking, leather welding gloves do not insulate as well as asbestos gloves. The latter may be a better choice when working for a long period of time around molten metal. Keep in mind, however, that asbestos gloves generally tend to be stiffer. There will be a loss of precise finger movement. As I mentioned earlier, there always seems to be some type of trade off between gloves that will totally protect your hands and gloves that facilitate finger movement. In the end, the choice is yours. You will probably end up having at least one pair of each.

APRONS

One last piece of clothing that is worth mentioning is an apron. A quality apron, designed specifically for use during welding, offers one very good alternative to traditional heavy duty clothing. Aprons are currently available in many different styles, materials and price ranges. To be effective, the apron should protect as much of the welder's front as possible while enabling the welder to have free movement of arms and body.

Leather and asbestos are probably the two most common materials used to make welding aprons. Both materials are quite effective at reflecting heat and deflecting hot sparks while cutting or welding. Look at the aprons offered at your local welding supply house. You quite possibly will find one that is suitable for your welding protection needs. Obviously all of the care and use requirements of clothing mentioned earlier apply to aprons designed for welding (Fig. 14-14).

FIRES AND FIRE EQUIPMENT

Because the possibility of a fire always exists for the welder, he should have some means of putting a fire out if one should start.

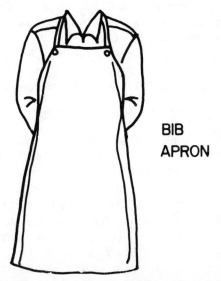

BIB
APRON

Fig. 14-14. A bib apron provides adequate protection of clothing.

Before you can put out a fire, you must know a few things about fires in general. The old military saying "know your enemy" is one of the most effective ways of dealing with fires. As a way of making you more aware of fires in general, we need to discuss the basic components of fires.

Fuel, Oxygen and Heat

Before a fire can take place, three things are needed: fuel, oxygen and heat. If any of these three components are removed, a fire cannot continue to burn. Whenever a fire breaks out, you must make a quick decision as to which of the three parts of that fire can be eliminated most easily, thus eliminating the fire. The three obvious means of doing this are removing the fuel (which can often be near impossible), cooling the material which is burning or removing the oxygen by smothering.

All fires have two things in common, heat and oxygen, and these can be an aid to you in fighting any fire. The third component of a fire, fuel, can vary from one fire to another. It is because of the possible variation of fuel in any given fire that all fires are grouped into four possible types: Classes A, B, C or D. You should know about each class of fire in order to determine the best way to extinguish any particular fire.

Class A Fires

Class A fires are those in which wood, paper, clothing and similar materials are burning. Generally speaking, water is the best and most effective way to fight a Class A fire. Water will penetrate into the burning material and cool it. Water will also, at least in part, seal the material from the oxygen in the atmosphere, although to a much lesser extent than other fire fighting materials.

Class B Fires

Class B fires consist of flammable liquids such as oil, gasoline, paint and similar materials. As a rule, once the flame is extinguished in a Class B fire, the fire will go out. You must, in effect, remove the oxygen or separate the burning material from oxygen to extinguish a Class B fire. Only the vapor on top of the liquid surface is burning and not the material itself in a Class B fire. For example, diesel fuel burns on the surface as the vapors evaporate into the surrounding atmosphere. The fuel itself does not contain oxygen. Separate these fumes from the oxygen in the atmosphere and the fire will go out.

The most effective means of fighting a Class B fire is, of course, to smother the fire, thereby separating the vapor, which is burning, from the fuel or burning material, which is not burning. Foam, powder, sand and other types of nonflammable materials are most effective because of the heavy smothering action. Once the burning material is separated from the oxygen in the atmosphere, the fire will go out almost immediately.

In no case should water ever be used to try and extinguish a Class B fire. Most flammable liquids will float on top of water and continue to burn. Water will not extinguish the fire. When water is thrown on a Class B fire, it may actually float the flames and fuel over a large area which could dramatically spread the fire and further prevent putting it out.

Class C Fires

Class C fires are electrical in nature. This class of fires includes electrical wiring, motors, switches, machinery and appliances. A smothering action is the best means of fighting a Class C fire. For our purposes, that means a nonconductive material, generally a special fire extinguisher. Because of the possibility of electrical shock, water should never be used in an attempt to extinguish an electrical fire.

Class D Fires

Class D fires are rather rare and almost nonexistent for the home welder. These fires are caused by burning metals such as magnesium shavings. The only way to put out a Class D fire is to use a specially designed fire extinguisher specifically labeled for Class D fires. If the home welder ever plans to work with combustible metals, he should have a fire extinguisher of this type handy.

Any fire will fall into at least one of the above classes of fires: A, B, C or D. Probably the best way to fight and extinguish a given fire is to use a fire extinguisher that has been designed to put out a specific class of fire.

Fire Extinguishers

As you may have guessed, there are fire extinguishers for each class of fire. There are also fire extinguishers which are are suitable for more than one class of fire. These will be clearly labeled. Probably the best choice for the home welder is a type generally referred to as multi-purpose or all-purpose fire extinguisher, often labeled for ABC fires (Fig. 14-15).

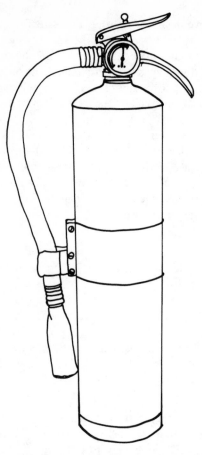

Fig. 14-15.   This fire extinguisher is designed to fight Class A, B and C fires.

The best types of fire extinguishers will have a pressure gauge at the top of the unit which will indicate the condition of that particular fire extinguisher. Some are more elaborate than others, but all that is really required is a gauge that will indicate whether or not the unit will work when needed. A gauge like this will give you valuable information at a glance, *recharge* or *operable*, for example (Fig. 14-16).

Even though your fire extinguisher may have a gauge, you should have it checked by a qualified repairman once a year. This will insure that the unit is, in fact, in good condition and that it will be an effective means of fighting a fire.

If you ever use a fire extinguisher, you should have it refilled, even if you only used it for a few seconds. What would be the point of having a fire extinguisher around your shop if it was only

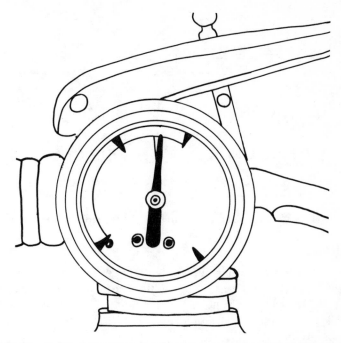

Fig. 14-16. A fire extinguisher should have a gauge that tells if the unit is operable or not.

half full or not properly charged? Your property and possibly your life may depend on your ability to quickly extinguish a fire. That is reason enough to make certain your fire extinguisher is in first class working order.

Sand and Water

In addition to a quality and multi-purpose fire extinguisher, you should also have a bucket of dry sand and a pail of water in your workshop or in the area where you are working. Sand can be used quite often to extinguish a Class B fire because it will quickly smother the flames. A bucket of water should only be used for putting out a Class A fire for those reasons previously mentioned. Nevertheless, both sand and water can be used to quickly extinguish certain types of fires and offer alternatives to your fire extinguisher (Fig. 14-17).

Obviously, your best defense against fires is to prevent them from happening. That means working only in safe surroundings. Since the next chapter covers many aspects of the home welding shop, you will have to turn there for more information on the best type of surroundings in which to weld or cut.

Fig. 14-17. Sand and water can be used to fight certain types of fires.

BURNS

Since we are discussing fires, it will be worthwhile to mention a few things about burns. To say the least, burns resulting from welding or cutting are not a pleasant topic for discussion. Nevertheless, the home welder should know what to do in the event of a serious burn.

Generally speaking, all burns can be grouped into three rather large categories. Since treatment of any one particular burn will depend on the nature of that burn, we will discuss all types. All burns fall into either first, second or third degree burn categories.

First Degree Burns

First degree burns result in red skin and possibly a small amount of pain. Sunburn is usually a first degree burn as is red skin, due to reflected heat from welding or cutting. Generally speaking, first degree burns are more uncomfortable than serious. The best treatment for first degree burns is to relieve the slight pain by applying a burn ointment to the area. If the burn covers a large area, see a doctor as soon as possible. Generally, the pain will subside in a few hours. If pain persists, seek medical help.

Second Degree Burns

Second degree burns are a bit more serious and are characterized by blisters. Treatment for second degree burns should *not* include covering the blisters with a burn ointment. You should prevent the blisters from breaking by covering the area with sterile gauze and holding it in place with a bandage. Second degree burns are serious and medical attention should be given by qualified personnel as soon as time permits. The greater the area of the second degree burn, the more potentially serious the problem.

Third Degree Burns

Third degree burns are the most serious type of all burns and are characterized by burned or charred flesh. Because of the nature of welding and cutting equipment, third degree burns are common. It will therefore be to your advantage to know what to do if this type of burn should happen to you or someone else in the welding shop. Know in advance that third degree burns which cover more than 50 percent of the body often result in a fatality. Therefore, quick action on your part may make a difference.

There are a few cardinal rules that should be followed when dealing with third degree burns. They are as follows:

- Do not move the person unless absolutely necessary.
- Do not apply burn creams or salves to cover the burn.
- Do not remove burned clothing. Leave it in place to be removed by qualified medical personnel.
- Treat for shock by covering the victim with a blanket or other suitable material. Lay the person down with feet elevated, if practical.
- Seek medical attention as quickly as possible, either by rushing the victim to the hospital or calling in emergency medical assistance.

All burns require attention as they are all potentially serious. It is important to keep in mind that quick action on your part may make a difference, so act accordingly.

HANDLING WELDING EQUIPMENT

The equipment used in welding and cutting is potentially dangerous and should always be treated with the respect that it is due. Failure to do so may result in mishap. Many points for safe operation can be found throughout this book. The majority of the remainder of this chapter will summarize those practices which should be standard operating procedures for anyone working around or with welding equipment.

Oxygen and acetylene cylinders, as well as other fuel gas containers have properties that can cause serious accidents, injuries and even death if proper precautions and safety practices are not followed. It is therefore very important that you become aware of these safety precautions. Make them a part of your welding habits.

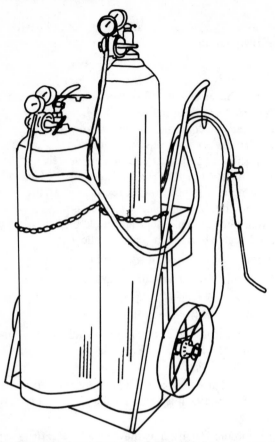

Fig. 14-18. Oxygen and acetylene cylinders should always be securely fastened so there is no chance of falling. They can be placed on a hand truck.

Because both oxygen and acetylene are in pressurized containers, they must be handled with respect. It is very important that these cylinders be securely held in an upright position at all times. Probably the best way of doing this is to chain both cylinders to a special hand truck designed to hold oxygen and fuel gas cylinders. These hand carts are available where you purchase your oxygen and acetylene (Fig. 14-18).

If the tanks are to remain in one location, make certain that both are secure by chaining them to an immovable object like an interior wall or column. Never allow either type of cylinder to stand by itself for any period of time. This will be your best insurance against accidental tipping of the cylinder.

Keep all organic materials as well as flammable substances away from both oxygen and acetylene cylinders. Remember that even

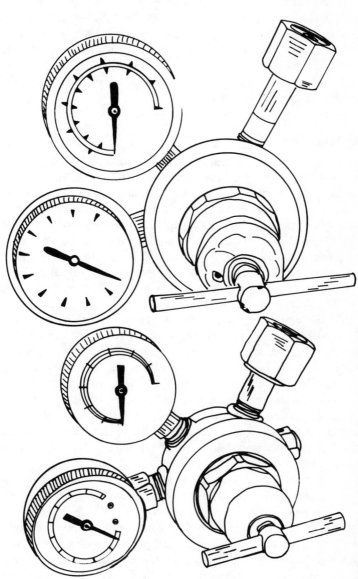

Fig. 14-19. Always use the proper regulator for oxygen and fuel gas.

normal soot and dirt can constitute a combustion hazard around these gases.

Never use either oxygen or acetylene (or other fuel gases) without first attaching an approved regulator to the tank. A full tank of oxygen will have internal pressure of about 2,200 pounds. Full acetylene cylinders will have an internal pressure of up to 400 pounds. It is therefore extremely important that only a suitable regulator be used to reduce these pressures to recommended working pressures in the lines (Fig. 14-19).

Handling Oxygen

In addition to the general rules for safe handling of both oxygen and acetylene, there are other recommendations that apply to the specific gas. Never allow oxygen to come in contact with oil, grease, kerosene, paint, tar, coal dust or dirt. Oxygen supports combustion and can burn violently in the presence of these and other flammable substances. You should never handle oxygen or equipment that may come in contact with oxygen like the regulator, with dirty hands, grease-stained gloves or other soiled clothing.

Before attaching a regulator to the oxygen cylinder valve, wipe the fitting and threads with a clean, dry cloth (Fig. 14-20). You should also crack the valve on the oxygen cylinder to blow out any dirt or other foreign matter that may have found its way into the valve. When cracking the oxygen cylinder valve, do so quickly. Position your body so that the oxygen in the tank will not spray on you. Stand to the side of the tank valve (Fig. 14-21).

Never use oxygen for ventilation or as a substitute for compressed air. You should never blow the dust off anything with oxygen. Cleaning your clothes off with a blast of oxygen and then lighting up a smoke is one of the quickest ways to end up in the hospital.

Always refer to oxygen as "oxygen" and never air. This will avoid the possibility of confusing oxygen used in the welding process with compressed air.

Never use oxygen for other than its intended purpose. Oxygen that is designed to be used for medical purposes such as respiratory ailments will have U.S.P. stamped on the cylinder. All other types of oxygen are for industrial uses only and should not be used for oxygen therapy.

When transporting oxygen, do so only with the container in an upright position in an open vehicle. Never carry an oxygen cylinder laying on its side in the closed trunk of an automobile.

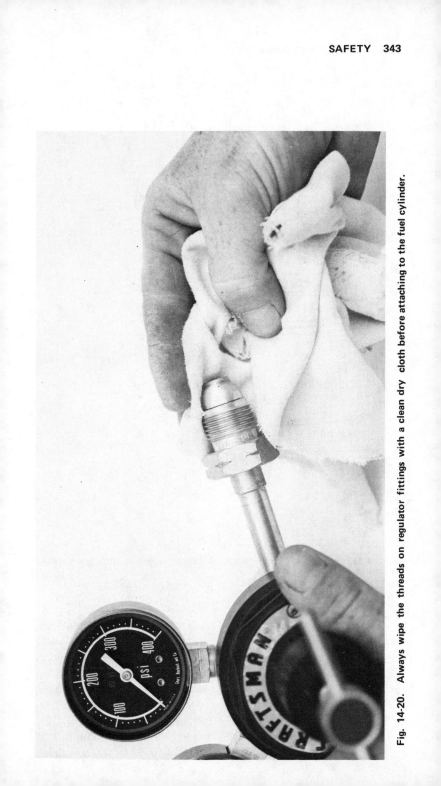

Fig. 14-20. Always wipe the threads on regulator fittings with a clean dry cloth before attaching to the fuel cylinder.

Fig. 14-21. Stand to the side when you crack an oxygen or acetylene cylinder valve.

Handling Acetylene

Acetylene cylinders are equally dangerous. All of the guidelines for safe operation and use of oxygen cylinders should be followed for this fuel gas as well. In addition, you may recall that all acetylene cylinders have a safety plug in either the top or bottom of the cylinder. This safety plug is designed to melt at any temperature above 212 degrees Fahrenheit. It is a safety feature intended to reduce the hazards of an explosion in the event of a fire (Fig. 14-22). Because this safety plug has such a low melting point, you should never work with an open flame around a cylinder of acetylene.

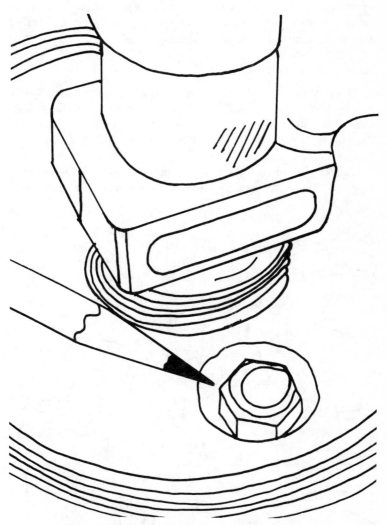

Fig. 14-22. The safety plug in the top of an acetylene tank.

For the same reason, you should never pour boiling water on top of an acetylene valve in an effort to thaw out a frozen valve. This can happen when the acetylene cylinder has been stored in an unheated room or outdoors during cold weather. At all times you should protect the fusible safety plug in acetylene cylinders from damage. Store, transport and use acetylene in a vertical position only.

REVIEW OF EQUIPMENT SETUP

When setting up welding equipment, namely oxygen and acetylene cylinders, follow those guidelines set down in Chapter 1. A quick review of those steps may further drive home the importance of proper handling of all welding equipment.

Oxygen Connections

Make certain that the fuel and oxygen cylinders are securely held in one place that will permit them to have a safe, happy life. Crack the valve of the oxygen cylinder before attaching the special oxygen regulator. Wipe the threads of the fittings of both the regulator and cylinder valve before fastening the regulator in place.

Tighten the regulator with a wrench designed for the purpose (Fig. 14-23). Undertightening will result in a leak, and overtighten-

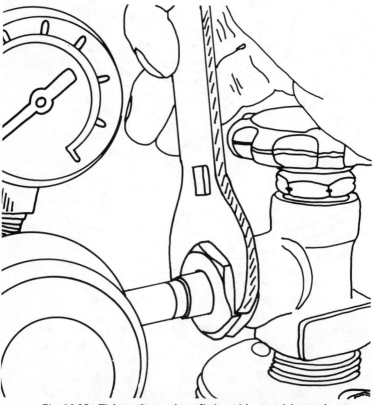

Fig. 14-23. Tighten the regulator fitting with a special wrench.

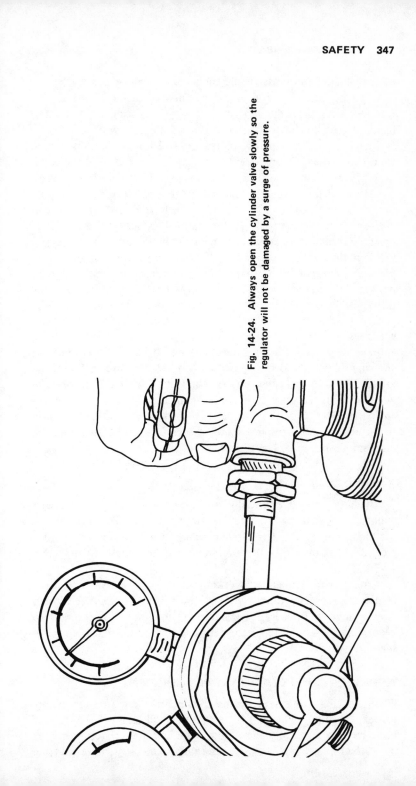

Fig. 14-24. Always open the cylinder valve slowly so the regulator will not be damaged by a surge of pressure.

ing may cause the brass fitting to become strained and weakened.

Connect one end of the oxygen hose (usually green) to the regulator and the other end to the oxygen fitting on the torch handle. Snug up all connections with a suitable size wrench.

Check to make sure the oxygen control knob on the blowpipe handle is closed before opening the valve on top of the oxygen tank. In addition, open the regulator control lever until there is no resistance to the turning. The diaphragm is now open. Open the oxygen control valve on the cylinder slowly (Fig. 14-24). This will insure that the delicate oxygen regulator does not receive a surge of pressure from the cylinder. Remember that a full tank of oxygen will contain approximatey 2,200 pounds per square inch of internal pressure Once a bit of oxygen is in the regulator and the high pressure gauge registers the pressure in the tank, open the valve two turns. At this time, give a quick check to the regulator to make certain that it is not leaking. If you suspect it is, use the soapy water test.

In most cases a leak is a result of not tightening a connection enough. If you discover a leak, first turn off the valve on top of the tank. Then try snugging up the connection with the proper size wrench. Put pressure into the system and check the suspected area (Fig. 14-25). If the leak persists, there may be internal damage to the regulator or fitting. It should therefore not be used until the problem has been sufficiently corrected.

Acetylene Connections

The procedure is essentially the same for hooking up acetylene connections. Keep in mind, however, that all connections for acetylene are of the left-handed type and will be marked with a groove cut around the circumference of the brass fitting (Fig. 14-26).

After the recommended working pressures have been set on the regulator, open the torch handle momentarily to see that oxygen or acetylene is, in fact, flowing through the line. You should also keep one eye on the regulator dials and make sure that they move slightly, if at all.

WORKING WITH OXYACETYLENE EQUIPMENT

When working with oxyacetylene equipment, there are a few points to keep in mind. The first is that you should turn off the flow of both oxygen and acetylene at the cylinders if you stop work for more than a few minutes. The second is that if you are shutting

Fig. 14-25. Set the working pressure by turning the lever until the desired pressure is reached.

down the equipment for more than a half hour you should, in addition to turning off the flow of oxygen or fuel gas, also bleed the lines. Clear the acetylene line first by opening the control knob on the blowpipe handle. Watch the acetylene regulator when you do this. In a matter of moments both gauges, if so equipped, should register zero. Then turn off the acetylene control knob on the blowpipe handle. After you have cleared the line of acetylene, follow the same procedure for clearing the oxygen line (Fig. 14-27).

Obviously, when you are shutting down the oxyacetylene equipment, you should make certain that there is no open flame in the area. When working with oxyacetylene equipment, it is important to always have a clear path to the tanks or oxygen and fuel gas. In the event of mishap, one of the first things you should do is to turn off the flow of both oxygen and acetylene at the top of both tanks.

Another good practice when welding is to always make sure that the hoses carrying oxygen and acetylene do not lie under the work. A hot spark or piece of hot slag falling on a hose could easily burn its way through with drastic results.

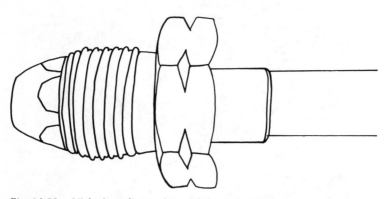

Fig. 14-26. All fuel gas fittings have left-handed threads and a notch around the fitting nut.

Store your welding equipment and tanks in a safe place. This equipment, for example, should never be left within easy reach of children. Both oxygen and acetylene cylinders come with a special valve protection cover. This cover should be in place over the valve whenever the regulator is not attached. If children have any possibility of being in the area where oxygen or acetylene is stored, the valve protection cap should be made secure and, at the very least, tighter than small hands can loosen (Fig. 14-28).

Hose, torch, cutting attachment and tips should all be stored in a safe place as well. A cabinet that will keep these parts dust free is best. It will go a long way in preventing costly repair bills as well as mishaps around the workshop.

WELDING HAZARDS

In summing up this chapter on safety, I think it will be helpful to briefly outline the hazards that are always present in welding. As a welder, you should know these hazards and continually strive to take them into consideration. This will be your best insurance against an unfortunate and possibly tragic experience.

- **Fire**. The possibility of fire always exists. You should therefore always have at least one means of extinguishing a fire if one should break out.

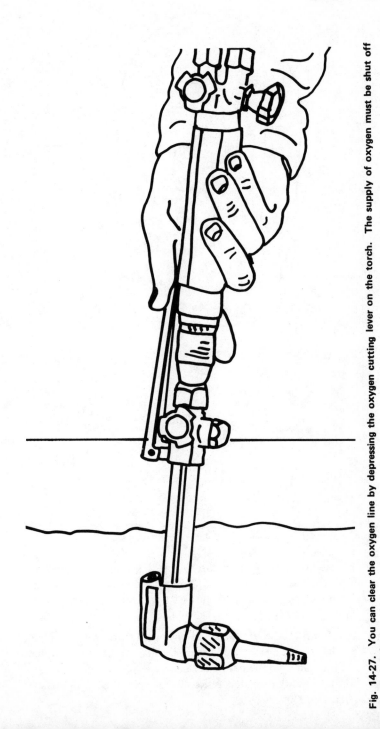

Fig. 14-27. You can clear the oxygen line by depressing the oxygen cutting lever on the torch. The supply of oxygen must be shut off at the tank first.

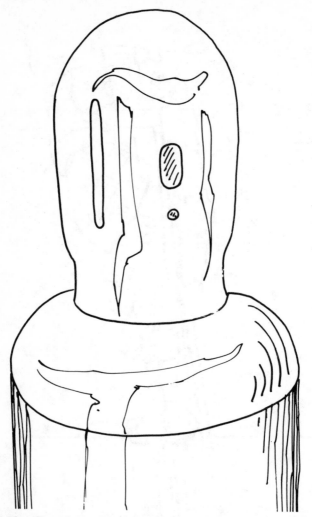

Fig. 14-28. The safety protection cap should always be on a cylinder of oxygen or acetylene whenever the regulator is not attached.

- **Burns.** Because of the nature of welding equipment, a serious burn can happen at any time. Always wear protective clothing to reduce the chances of burns to yourself or others in the shop.

- **Eye Injury.** Always wear tinted eye protection when welding, cutting or even simply adjusting the flame of your blowpipe. Additionally, you should wear clear eye protectors when performing welding related tasks such as grinding metal surfaces (Fig. 14-29).

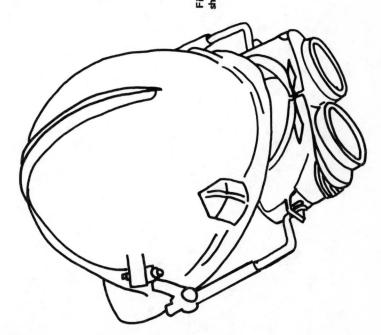

Fig. 14-29. Eye and head protection should always be worn when cutting or welding.

- **Explosions.** Always handle, transport and store oxygen and fuel gas cylinders with the respect due them. Check all connections to make certain that no fuel gas or oxygen is leaking into the atmosphere. Exercise care around other containers with flammable materials.

- **Harmful Or Poisonous Gases.** Welding or cutting some types of metals like galvanized steel may result in a harmful gas. Make certain that the welding area is adequately ventilated at all times. Wear a respirator over your mouth and nose whenever grinding or welding potentially dangerous metals.

- **Mishaps.** A danger always exists when working in unsafe surroundings or as a result of unsafe work habits.

- **Failure To Exercise Caution And Common Sense.** Think before you act and you will eliminate 95 percent of the danger associated with welding and cutting operations.

A wise welder knows and practices safe work habits. He is constantly aware of the dangers involved in welding and does everything in his power to make welding and cutting a safe and enjoyable means of accomplishing various metal joining tasks.

Chapter 15
The Home Workshop

In addition to oxyacetylene welding equipment — blowpipe, cutting attachment, safety equipment, hoses, tanks of fuel gas and oxygen and related accessories — the home welder requires a safe space to work. If you think that you can take care of all of your welding projects in the back of your garage or in your basement, you are flirting with danger. Welding, because of the inherent risks, requires that the work always be done in a special area. After all, we are dealing with controlled fire and very hot metal when welding or cutting. If special precautions are not taken, this can lead to an unfortunate experience at the very least.

OUTDOOR WORKSHOP ADVANTAGES

When just beginning, chances are very good that you will be practicing cutting and welding outdoors. This makes sense on a number of levels. Even though there are some drawbacks to working outside, you should find an outdoor workshop more than adequate for your practicing needs.

Ventilation

Probably the greatest advantage of working outdoors is that there is always plenty of ventilation. Even on windless days, smoke and fumes from welding and cutting will quickly drift up and away from the area. As you may have already discovered, the resultant smoke from welding, even though nontoxic, can be upsetting to the system. It is therefore a good practice to inhale as little smoke or fumes as possible.

Equipment Setup

Setting up equipment, as well as shutting down the equipment, can be safely done outdoors as well. For example, when you crack the oxygen or acetylene valves prior to attaching the regulators, these gases will quickly disperse into the atmosphere and pose little danger. The same "clearing of the air" also takes place quickly when bleeding the oxygen and acetylene lines while the equipment is shut down. These gases will simply drift away. When you are working indoors, you must make certain that these gases are exhausted from the area by some artificial means like a ventilator fan. Later in this chapter we will discuss the various means of insuring fresh air, and plenty of it, in an indoor welding workshop.

Minimal Cost

Another advantage of an outdoor workshop for welding is that the cost of setting up such a shop will be, in most cases, quite minimal. In addition to the necessary oxyacetylene welding and cutting equipment, you will also require a worktable. Since a worktable is one of the basic components of any welding shop as well as a real aid to working with oxyacetylene equipment, you will want to build one as soon as you have mastered some of the basic cutting and welding techniques discussed earlier in this book.

WELDING TABLE

To build a simple welding table, you will need 20 feet of 2 x 2-inch angle iron, 20 feet of 1-inch wide steel strap and 24 firebricks 8 x 4 x 2 inches. See Table 15-1.

The dimensions for a standard welding table are given in Fig. 15-1. Begin by cutting the angle iron into four pieces, each 28 inches long. Then cut the remainder of the angle iron into two 24-inch and two 32-inch long pieces. The angle iron (actually angle steel) comprises both the legs and the frame for the top of this welding table. You must cut precisely for best results.

Assembly

Begin assembling the table by laying out the top pieces and welding them together (two 24 and two 32-inch long pieces). You will

Table 15-1. Materials List for the Welding Table.

4 — 28" pieces 2x2 angle iron
2 — 24" pieces 2x2 angle iron
2 — 32" pieces 2x2 angle iron
4 — 24" pieces 1" steel strap
5 — 32" pieces 1" steel strap
24 — fire bricks 8x4x2"

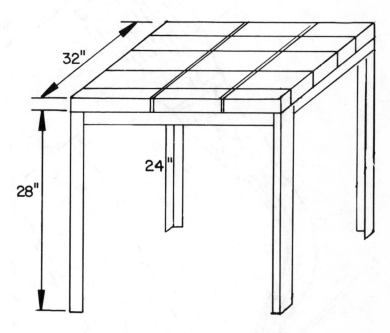

Fig. 15-1. Dimensions of the welding table.

be welding a rectangular frame that when finished will measure 24 inches wide by 32 inches long. After this frame has been welded, you must then weld the legs, one on each corner. Now you will have a basic frame table. All that remains is to weld the steel strap, four 24-inch long pieces and five 32-inch long pieces, inside the frame top. Placement is important because the steel strap provides support for the firebricks which will be the working surface of this table.

It will probably be easiest to begin by welding the 24-inch long steel strap first. Space the strap at 8-inch intervals and tack weld them in place. Next, the 32-inch long strap is either laid on top of the cross members or woven through them for additional strength. Just the ends of these longer strips are tack welded. You must place these straps at 4-inch intervals so the edges of the firebrick will lie on them. If you are in doubt as to the placement of the steel strap,

Fig. 15-2. Tack weld the steel strap around the inside of the table frame. Proper spacing is very important.

simply lay a few firebricks into the frame to give you an idea of the exact placement (Fig. 15-2).

As you can see from Fig. 15-1, this worktable will have a finished height of 28 inches. If you are taller or shorter than average, you may want to build a table with a different height. To do this, simply adjust the length of the legs. It may be of interest to you to know that you will have some extra 2-inch angle iron if you buy the listed 20 feet. Actually you will have an excess of about 16 inches, so you will have little problem increasing the height of this table by about 3 inches with the materials you have on hand. Twenty-eight inches is, by the way, the standard height for a work surface, and most people find it about the best.

Modifications

Modifications are, of course, possible with the basic plans. Some additions you may want to include are roller type wheels for each of the legs. The addition of wheels will make the table that much more versatile as it can then be moved about quite easily.

You may also want to add additional angle iron midway between the top of the table and the floor or ground. Then you can add a shelf to hold scrap metal or other materials. Keep in mind, however, the potential fire danger of storing materials under a table that will be used for cutting and welding.

Another addition might be a 1/4-inch thick steel plate, 24 inches long and 8 to 12 inches wide. Attach this plate to one end of the worktable and you will increase its versatility by providing a solid and unbreakable work surface.

You can also attach hooks on one end of the tabletop edge. These can be used to hang various and often used tools such as a hammer, clamps and spark lighter. The point is to make your worktable useful to you as a welder. I would suggest that you build the basic table and use it for a period of time (Fig. 15-3). Then, after you have a clear picture of your work surface requirements, add personal touches to make your work easier and possibly more enjoyable.

Outdoor Workshop Sites

Your outdoor workshop can be almost anywhere around your home that space permits. You must have access to electricity for running a grinding wheel, drill and possibly other equipment. You will also find it much easier if you can set up your equipment and begin work quickly rather than making a half-dozen trips to the

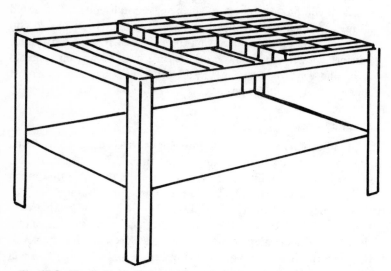

Fig. 15-3. The finished welding table made from angle steel and steel strap.

house or garage to move the required materials and tools out to the work area. In most cases, this will mean that your outdoor work space will be close to your house or garage. It will be helpful to discuss some of the possible sites for an outdoor workshop while pointing out some of the requirements of a good spot as well.

Carport

A *carport* can make an ideal outdoor workshop. You will be able to leave your welding equipment and tanks in one spot without having to worry about the elements of wind and water playing havoc with them. You will also be able to store scrap and unused metal out of the weather. Ventilation will be of little problem as most carports are open on at least two sides. You must make certain, however, that flammable materials are not in the area. In some cases, this will mean leaving your automobile and lawn mower in another location, as well as paints or other materials, when welding or cutting. If a concrete floor exists in your carport, it should be clean and free from oil or grease stains. If wooden walls exist in the carport, you should locate your welding table as far as possible from them. You can cover existing walls with asbestos board, commonly sold in sheets from 2 to 4 feet wide and from 4 to 8 feet long.

Cement or Stone Patio

If you are not fortunate enough to have a covered carport, which can be used as a welding shop, consider a cement or stone patio which is close to the house. Keep in mind, when using an uncovered area for welding projects, that you will be limited by good weather as to when you can work. You must also rig up some means of covering the oxygen and acetylene cylinders if they are to be left outside (Fig. 15-4). In some areas of the country you can get by

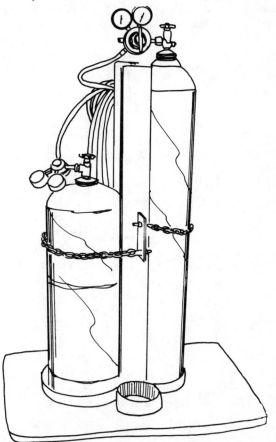

Fig. 15-4. When oxygen and acetylene tanks are stored outdoors, they must be chained to a solid object and covered when not in use.

quite nicely by covering the equipment with a plastic tarp when not in use. In other parts of the country, this may not be the best protection. In the Pacific Northwest, for example, inclement weather is quite common. You can also work on bare ground, if necessary, providing that all materials such as fallen leaves are removed from the area each time you plan on using the outdoor shop.

Outbuilding

If you live on a farm or ranch, you quite possibly may have an outbuilding that can be used for a welding shop. You may also find that a number of welding tasks need to be accomplished in places other than the welding shop. In cases such as these, you will find it very handy to have your oxygen and acetylene tanks on some type of portable hand truck (Fig. 15-5). Then you will be able to take the welding to where it is needed rather than the other way around.

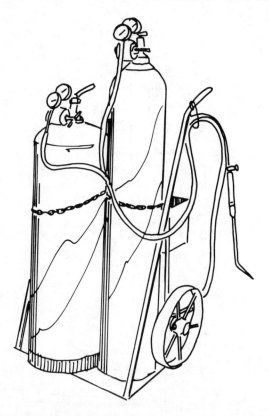

Fig. 15-5. A special hand truck is one of the best ways of transporting and storing oxygen and acetylene cylinders.

Whenever you work outside, which in many cases will be on bare earth, you should clear the area of any materials that might catch on fire during welding or cutting operations. If you are working during the dry part of the year, it may be a very good idea to wet the earth with a garden hose before beginning work. Keep the hose close to the area you are working in so you will be able to quickly extinguish any fire that might occur. Needless to say, other fire fighting equipment should be handy as well.

INDOOR WORKSHOP NEEDS

If outdoor space is not available or if you have suitable indoor space for welding, such as a garage, there are a number of things that you should take into consideration before any welding or cutting project is started. As I mentioned earlier, adequate ventilation is an important prerequisite of any welding area. When you are working indoors, you must make certain that the air in the space is fresh and moving. In most cases a common exhaust fan, the type found in most kitchens, will do the job quite nicely (Fig. 15-6). If you are

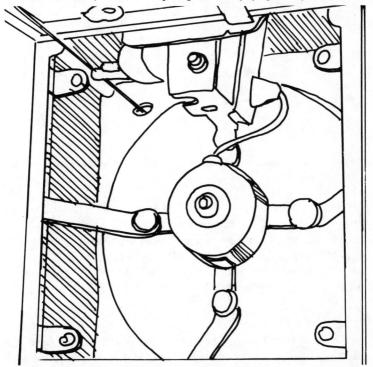

Fig. 15-6. You must install an exhaust fan in an indoor welding shop to remove the fumes from welding and cutting operations.

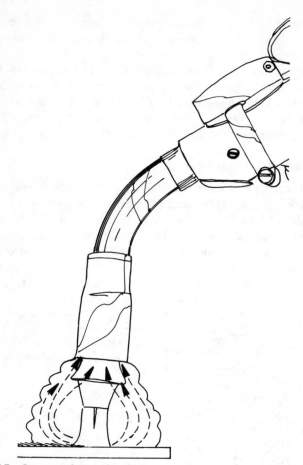

Fig. 15-7. One manufacturer's solution to the smoke problem when welding is a smoke exhaust gun system.

working in a large area, you may find it necessary to install a heavy duty, large capacity fan to ventilate the room. The point is that all smoke and fumes must be removed from the work area as quickly as possible (Fig. 15-7). You should also remember that fresh air has to come from the outside of the room. You may, in addition to installing an exhaust fan, also have to install a fresh air vent.

Fire Station

Your home workshop, whether indoors or out, should have adequate fire fighting equipment that is accessible and ready to use at all times. In most cases, it is best to designate one particular area as a *fire station.* Here you should keep a fully charged fire extinguisher, bucket of sand and a pail of clean water. The fire station area should never be used for anything like storage of scrap metal. There should always be a clear path to the equipment. That way, if you are ever in need of fire fighting equipment, you will know where to get it as quickly as possible. A fire station may quite possibly save you from the severe consequences of an out of control fire in your workshop.

Accessibility

Still another requirement of an indoor welding workshop is that you must have accessibility for delivery of supplies such as tanks of fuel gas and oxygen. If your welding shop is in a garage, there should be little problem in getting materials and supplies into the shop. If your welding shop is in your basement, you may find deliveries a real exercise in equipment manipulation. Consider easy access as one very important prerequisite of any site you are planning to use as a welding shop.

Lighting

Lighting is important in any workshop, and a welding shop is certainly no exception. It is a fallacy to think that you can work in a welding shop under a single light bulb. In addition to good lighting over your welding table, you must also have direct light over your bench grinder and other work areas, such as where you solder or where your bench vise is located.

Probably the best type of lighting for a welding shop is fluorescent (Fig. 15-8). Two 4-foot long tubes over the welding table should provide you with plenty of illumination. Fluorescent lighting

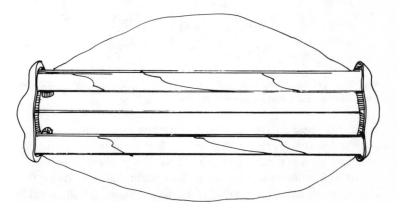

Fig. 15-8. A fluorescent lighting fixture will provide good general lighting in the work area and at a lower cost than incandescent lighting.

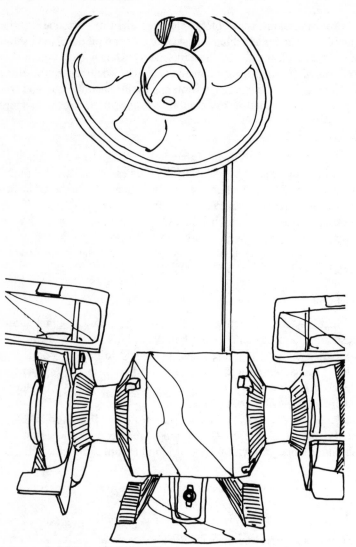

Fig. 15-9. It is important to provide adequate lighting over your grinding bench. This clamp-on light is inexpensive and does the job.

is also the cheapest type to run. For lighting in specialty areas, such as over the bench grinder, it is usually best to install a lighting fixture over each particular area (Fig. 15-9). Some possibilities include a clamp-on reflector lamp or small fluorescent fixtures. If you find that you will require a number of lighting fixtures, it will probably be best to install the fluorescent type as they generally run cooler and, as mentioned, are less expensive to operate in the long run.

While we are on the topic of electrical lighting, it may be a good time to mention that the indoor welding shop should have a number of electrical outlets. Various tasks will be easier to accomplish if you can plug equipment into any one of several electrical outlets. Considering that there are a few electrical tools which might be commonly used around the welding shop, you will want to make provisions for them. You should have a few outlets in addition to those already in use by stationary equipment or machinery.

You may also want to install a 220 volt electrical outlet in the welding shop. Most arc welding units require this higher voltage for operation.

Storage Space

Another consideration when planning a welding workshop and one that will become more apparent as you spend more time welding is that you will almost always require more space than you have. Such is life. Just how much space you really need will depend on your work habits and even the type of work you plan on doing. In all probability you will be limited to a garage or other outbuilding plus the area outside the shop. Over a period of time you will undoubtedly accumulate metals, both new and scrap, and will want to store them out of the elements. You will also probably find yourself with a number of tools and machines that are necessary for various metal cleaning or joining tasks.

The obvious solution to "all that junk" as my wife is fond of saying is to build storage units in one area of the workshop. Sturdy shelves can be used for storing small pieces of metal. Long pieces, such as angle iron or strap, are best stored flat on some type of rack system. Plate and sheet metals can easily be stored standing up in some type of vertical shelving system. Odd sizes and shapes of metal are usually a problem to store. Often a metal garbage can, with three or four caster type wheels welded to the bottom, will take care of the problem. In all cases it is always best to store metal off the ground, even if a concrete floor exists, to prevent rust on the metal (Fig. 15-10).

Chances are that most of the metal you work with will be steel in one alloy or another. You may also have some nonferrous metals around the shop as well. It will be better, in the long run, to store all similar metals together. Some type of labeling system will be helpful in identifying these metals.

If your workshop is a small one, you will probably be forced to store much of your metal outdoors. To keep the metal from rusting,

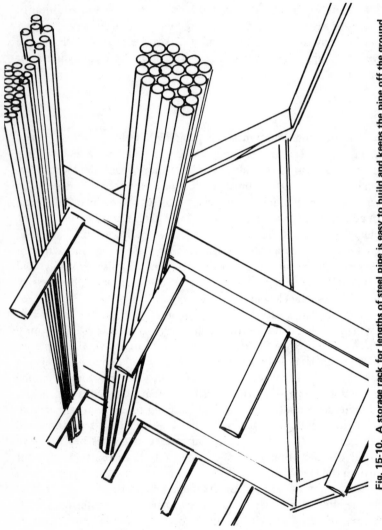

Fig. 15-10. A storage rack for lengths of steel pipe is easy to build and keeps the pipe off the ground.

you must cover it with plastic or a canvas tarpaulin. Keep in mind that the best place to store metal is in some type of closed shed. The more serious you become about working with metals, the closer you will get to building a storage shed for your supply of metal.

ACCESSORIES

There are any number of useful projects that you can build around your welding shop. In addition to having specific areas for storing metals, you will also need storage cabinets for tools and accessories. Since cabinet building is really beyond the scope of this book, I will not deal with these woodworking projects here.

Mobile Cart

One project that I will briefly discuss, however, is a storage unit for welding and brazing rods. Plans for a basic unit are given in Fig. 15-11. This unit can be made into a mobile cart by simply adding four caster type wheels to the bottom. You will find such a cart quite handy for keeping your rods stored neatly as well as keeping them handy when you need them.

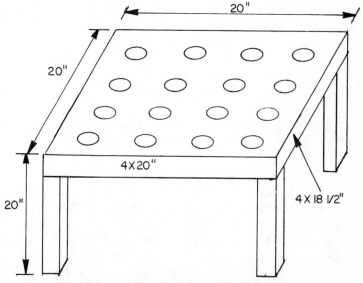

Fig. 15-11. A welding and brazing rod holder table.

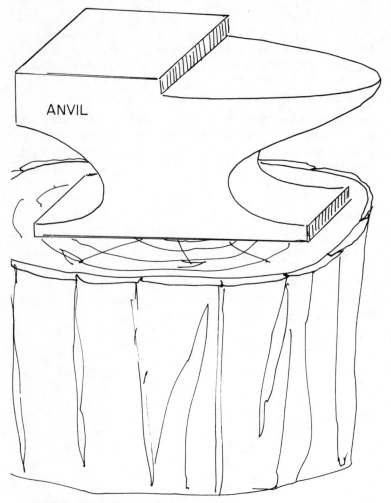

ANVIL

TREE STUMP OR LOG

Fig. 15-12. An anvil is handy for many metalworking projects. One can be made from a 16-inch long piece of railroad rail.

Anvil

Another useful accessory for your welding shop is some type of *anvil*. An anvil is quite handy for bending or flattening small work with a hammer. You can make one simply by cutting one end of a piece of scrap steel rail in the shape of a traditional anvil. A piece of rail about 16 inches long will be more than adequate for your needs (Fig. 15-12). After the basic shape has been cut out of one end of the rail, pierce cut four holes on the bottom with the cutting torch. Use these holes for fastening the anvil to a solid surface. One very good base for your home anvil is a section of tree about 18 to 24 inches in diameter and about 20 inches high. The log will absorb most of the shock when hammering and will also help to deaden the noise as well. Finish off your anvil with a portable grinding wheel or disc sander.

LOCAL CODES

One last thing to consider about your home welding workshop is the local codes. In some resdential areas, a commercial welding shop is not permitted. It will be in your best interests to check with a suitable governing body to see if any such restrictions exist in your area.

As a side note about restrictions in certain areas, you should know that if you call your shop a studio, rather than a welding shop, you may find that the same restrictions do not apply. In many instances your workshop will be a studio in the sense that it will be a place where you can express yourself through the art of metalworking.

SUMMARY OF WORKSHOP REQUIREMENTS

Choosing a location for your welding shop is a major consideration. If you limit yourself to small projects and only work occasionally, you will obviously be able to work outdoors. However, if you are planning to devote a fair share of your time to welding and cutting, you will need a space that can accommodate your needs. In an effort to help you decide if a particular area is suitable for general welding and cutting needs, consider the following eight requirements of a safe and useful welding workshop.

- **Ventilation.** Some means of exhausting smoke and gases must exist or be installed in the area. This exhaust system must have the capability of quickly clearing the air in the workshop as well as supplying adequate fresh air.

- **Fireproof Construction.** The walls, floor and ceiling of the welding shop must be covered with a noncombustible material to prevent fire.

- **Work Areas.** A good welding shop will have different areas for specific metal working tasks. These might include a cutting and welding table, soldering workbench, metal grinding area and possibly a separate area for finishing or painting metal.

- **Lighting**. Before you can expect to work in a workshop, you must provide adequate lighting.

- **Fire Station**. Every welding workshop should have fire fighting equipment in a central location.

- **Storage**. Welding, cutting and related tasks such as grinding can be dusty affairs. A good welding shop will have storage facilities not only for metals but for tools and equipment as well.

- **Access**. An effective welding shop will have wide doorways which permit easy movement of equipment and supplies.

- **Cleanliness**. A safe welding workshop is easy to keep clean and should be kept that way at all times. This will do a lot to prevent fire and also make for pleasant working conditions.

In the final analysis, you will make the decision as to the best location for your welding shop. Obviously, it will be to your advantage to choose a location that is both adequate for your needs and enables you to safely practice the art of welding.

Chapter 16

BernzOmatic Oxygen and Propane Torch Uses

A BernzOmatic oxygen torch will be a great addition to your home workshop. This torch produces a hotter flame than the conventional propane torch. As mentioned briefly in Chapter 7, propane or Mapp gas instead of acetylene is used together with oxygen. This chapter will describe some tasks that can be undertaken with BernzOmatic and propane torches. First, though, an examination of BernzOmatic torch operating procedures is in order.

BERNZOMATIC TORCH OPERATION

The first step is to check and see that the oxygen and fuel valves are closed. The oxygen cylinder, which has a left-hand thread, must be screwed into the opening closest to the torch tip (Fig. 16-1). Insert

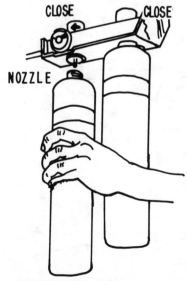

Fig. 16-1. Both valves must be closed before inserting the cylinders.

the propane or Mapp gas cylinder into the opening closest to the handle. This cylinder has a right-hand thread. You should know that Mapp gas is more costly than propane gas, but that disadvantage is negated by the fact that the former requires less time to heat up. Thus, labor time is reduced.

The fuel cylinder valve should be cracked very slightly, about one-sixteenth of a turn. Use a spark lighter, held at a slight angle, to ignite the flowing gas. Adjust the valve to produce a 4-inch flame. Make sure that the nozzle and flame are in contact with each other. You should then twist the oxygen valve to the left (counterclockwise) to generate a flame which is light blue in color. The best flame to use in welding is one with a light blue outer flame and a darker blue inner flame about 1/4 inch in length (Fig. 16-2). Decrease oxygen flow and increase Mapp gas flow to produce a flame of lower temperature.

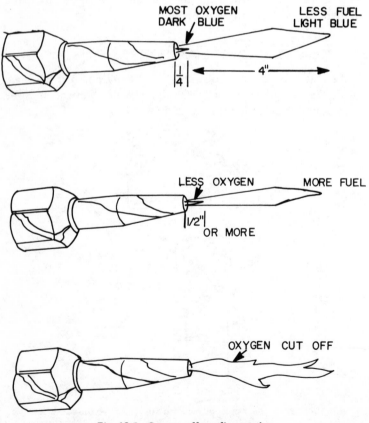

Fig. 16-2. Oxygen affects flame color.

When the oxygen supply has been exhausted or cut off, a long, yellow flame results. If the fuel cylinder valve is closed or if the cylinder itself is empty, there will be no flame. Close the fuel valve slightly to eliminate a flame which jumps off the tip. Always hold the torch in an upright manner and avoid excessive tilting.

You can conserve oxygen and fuel by not opening the valves as far as they will go. Most repair tasks can be performed without wide-open valves. When you have cleaned and clamped properly the metals to be joined, you are ready to begin welding with the BernzOmatic torch.

WELDING THE WORK

After lighting the torch, work to get a flame with a 4-inch long blue tip. The torch should be held to the work at about a 45 degree angle (Fig. 16-3). Position the tip in the course that the weld will

Fig. 16-3. Hold the torch at a 45 degree angle to the work.

take. Move the torch in whatever way works best for you. Either forehand or backhand welding can be used, and these styles are covered in Chapter 9.

Some welding can be done without utilizing a filler rod. Place the work so that you can begin at the right, assuming you are right-handed. The torch tip should point to the left at approximately a 45 degree angle to the work. The flame should be transported in back and forth 1/4-inch movements over the work area. In a zigzag manner, move the torch leftward as the metal commences melting. You will burn a hole in the metal if the torch is moved at a snail's pace. When you're almost finished, lift the torch to prevent burning of an unwanted hole. A weak weld is the result if the torch is moved too rapidly.

Occasionally the welding rod will adhere to the work. To loosen the rod, play the flame right at the rod's tip. Do not attempt to pull it loose. The rod sticks when it is not kept in the middle of the metal puddle. The puddle edges are close to the "freezing" point and often grasp the rod as it cools down. Remember that the puddle of metal will be hardening as you are attempting to loosen the rod. You can create a fresh puddle by reheating the metal.

The basics of welding with propane or Mapp gas are essentially the same as the principles involved with oxyacetylene welding. It is best to practice on scrap metal having the same characteristics as the material which you will be welding. As in oxyacetylene welding, safety is of prime importance. Keep fire fighting equipment on hand, wear proper clothing and gear and do not breathe resultant fumes.

REPAIRING TOOLS AND FRAMES

It's tough to keep weeds out of your garden when cultivator and rake teeth are loose or broken. With careful brazing, though, your damaged tool will be as good as new. The best brazing rods to use are the nickel-silver or the flux-coated bronze ones. Use coarse grade sandpaper and a wire brush to clean the break location. See Fig. 16-4 for the proper clamping procedure.

The sections to be joined must be heated to a cherry-red state. The rod should flow in an uninhibited manner when it comes into contact with the heated metal. No parts should be moved about. Let the joint cool and then test the quality of your work. If every-thing is okay, get back to weeding that garden!

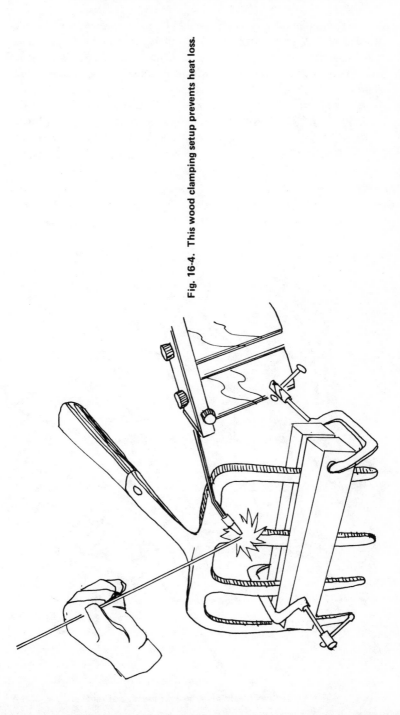

Fig. 16-4.   This wood clamping setup prevents heat loss.

Many household implements and kids' playthings are constructed around hollow tubular frames. These items include bicycles, barbecue grills, snow blowers, lawn mowers and swing sets. When rust sets in and weakens the framework, it won't take much pressure for a break to occur.

You can solve this problem by covering the affected section with two 6-inch pieces of steel conduit, which electricians use for wiring. Wrap each part of the tubing around the frame. You may have to use a hammer to get a perfect fit. Smack the hammer on the edges

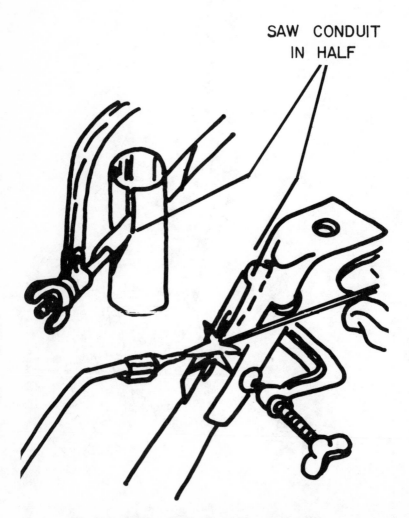

SAW CONDUIT
IN HALF

Fig. 16-5. Steel conduit can be used for various repairs.

if the fit is not tight enough. When the tubing is in place, use steel wool or sandpaper to reveal bare metal. Both the tubing insides and the frame's outside must be thoroughly clean. Then use a BernzOmatic torch to do the necessary welding or brazing (Fig. 16-5).

You can also use the BernzOmatic torch to weld a broken basketball hoop. Remove the hoop from the stand and clamp it in place to eliminate movement (Fig. 16-6). Use a nickel-silver or bronze

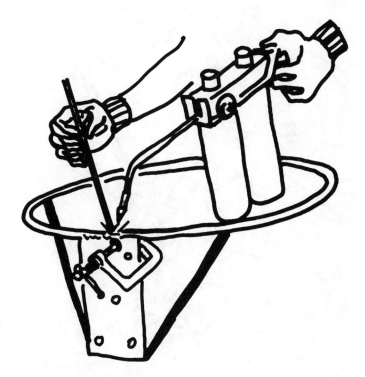

Fig. 16-6. You should remove the hoop from the stand to do the necessary welding.

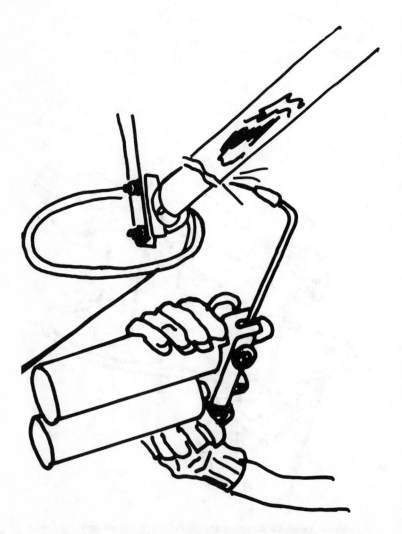

Fig. 16-7. Use caution when undertaking flame cutting.

flux-coated rod for a filler. After the weld has cooled, clean it and apply black enamel to the whole hoop assembly.

AUTOMOBILE MUFFLER WORK

If it's time for a new muffler, a BernzOmatic torch can aid you in removing the old one. You'll need to raise the two tires on the muffler side and place some bricks or blocks under each wheel. Always remember to put the car in park or in gear and set the brake. Place a block on both sides of the wheels on the surface. Now you're ready to make your way under the vehicle.

Adjust the torch to a pencil-like flame and apply heat to all unyielding nuts. Do this for about a minute. The object here is to enlarge the nut and loosen it from the bolt (Fig. 16-7). Position yourself so that you won't be the recipient of dripping molten metal. Keep the flame away from the gas line and tank. You don't want to risk a deadly explosion. If the bolts persist in being stubborn, you can use the torch to cut them.

It may be necessary to use the torch to cut the muffler free. Cut as nearly as you can to the muffler (Fig. 16-8). You should adjust valves to get a sharp edge to the orange flame and a sharp

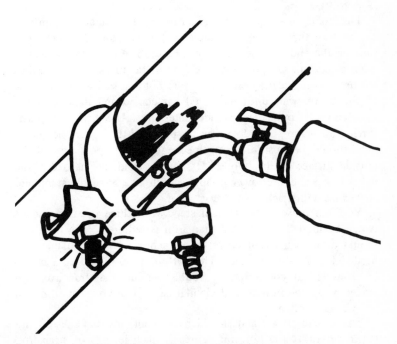

Fig. 16-8. Apply heat to help loosen nuts.

point to the blue one. The right length for the blue flame is about 1/4 inch. Apply the tip of the blue flame to the area to be removed. When the flame turns yellow-white, the fuel supply should be turned off. If the oxygen cylinder runs dry or the torch is moved too quickly, the metal will stop cutting. Work carefully and you should have no major difficulties. Then simply install the new muffler.

WORKING WITH WROUGHT IRON

Use the BernzOmatic torch to make plant stands constructed from wrought iron and strap iron. You will need 1/8 x 1/2-inch strap iron. Try making a three-legged stand which can hold a flower pot container. First, you will need to construct the two hoops or circles for the stand. Make certain that the upper hoop's inner diameter is 1/2 inch narrower in diameter than the container's outside diameter. You will want the top part of the container to fit closely against the upper hoop. The hoop keeps the container from dropping through the stand. The lower hoop should be big enough across to enable the flower pot container's bottom to pass easily through. When the hoops are bent into true circles, use C-clamps to hold them in place and weld the two hoops' ends (Fig. 16-9).

This stand requires three legs. For inches of metal comprise a curved foot for each leg. Completed legs will be 13 inches long. Curving the ends of the legs is not hard. The procedure does not require oxygen. Apply heat until the metal turns red-hot. Pliers are used to make the bend. If you're not satisfied with the shape of the curve, wait until the metal has cooled and then bend the material until you achieve an adequate curve. Use the BernzOmatic torch with Mapp gas and oxygen to spot weld the legs to the hoops. Legs should be positioned at three equidistant spots around the outside surface of the hoops. Of course, welds are needed at the upper hoop and the lower hoop for each leg. Sand the stand and coat twice with black enamel.

You can make a taller plant stand with three rings instead of two. Again, 1/8 x 1/2-inch strap iron is required. The top hoop should have an inside diameter of 9 inches. A 7-inch diameter is needed for the middle ring. The bottom hoop should be 3 inches in diameter. If you're having trouble bending the strap iron into circles, wrap the iron around circular cans. Use your torch to join hoop ends together.

The legs for this stand should have small curves at the top and larger curves at the bottom. Three 38-inch lengths of strap iron

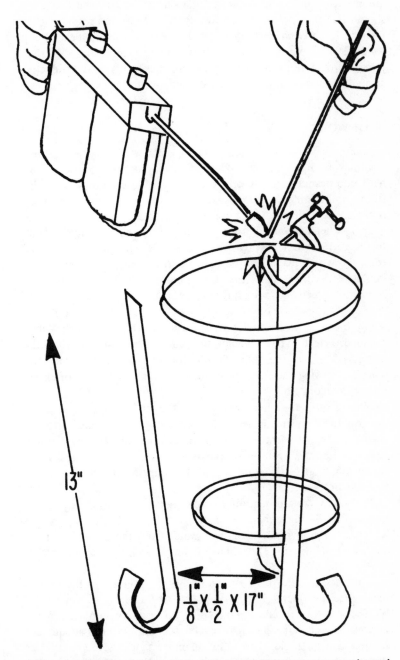

Fig. 16-9. Diagram of the two-ring plant stand. Make sure parts are clamped when welding.

are needed for the legs. After the curves are formed, each leg should be 30 inches long from the floor to the stand's upper hoop. Use the torch to make the curves. Use clamps and weld each hoop to the legs. Note in Fig. 16-10 that the legs bend outward as they join with the bottom ring. You can make this cold bend after assembly. Sand completely and slap on two coats of black enamel.

ROOF WORK

A common problem for homeowners today is sagging rain gutters, resulting in large amounts of water flowing over the gutters. Perhaps you have a gutter which is anchored by a type of strap. This strap may be soldered or fastened by rivets to the gutter. The strap's other end is nailed to the roof and concealed by shingles. If the strap has been loosened from the rivet, you should solder it back to the gutter. Remove any soggy leaves or other unwanted material from the work area. Sand the strap's bottom and the area where the joining will occur. Place the strap around the gutter and clamp it. Now you're ready for soldering (Fig. 16-11).

You can use a conventional propane gas torch for this job. Apply muriatic acid, which acts as a flux, to the area which will be soldered. Heat the far side of the joint. Apply solder when the metal is hot. If the solder begins to flow, add more solder and continue moving the flame until the joint is solid. Get rid of excess flux with an old cloth.

You can use a level to determine if the gutter slopes toward the downspout. Lacking a level, merely dump some water into the gutter and see if it travels toward the downspout. If not, you will have to alter the position of the strap. Take out the nails which fasten the strap to the roof. Move the strap so that the gutter will have the proper slant.

Perhaps a certain section of your rain gutter drips incessantly hours after the shower has stopped. It is probable that a joint has loosened. Remove all debris and water from the gutter and sand completely (Fig. 16-12). Then put on some flux. Use bar solder for this job.

Now it's time to heat the work. When you apply the solder, make sure the metal is plenty hot so the solder will run freely. Wrinkling solder means the metal is not yet hot enough. When you've finished, take pains to insure that the gutter is adequately supported near the joint area. For aluminum gutters, use the proper flux and solder. An acid core solder should be used on galvanized steel gutters.

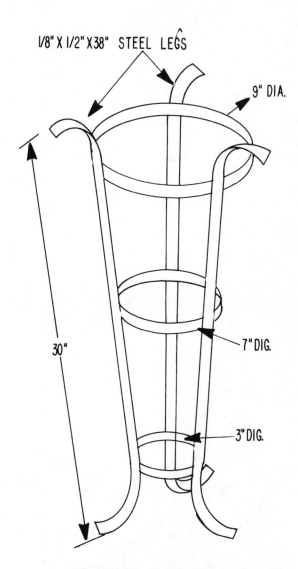

1/8" X 1/2" X 38" STEEL LEGS

9" DIA.

30"

7" DIG.

3" DIG.

THE HIGH BOY PLANT STAND IS SIMILAR
TO THE LOWBOY STAND EXCEPT, OF COURSE,
FOR ITS GREATER HEIGHT AND THE EXTRA
RING FOR ADDED STABILITY.

Fig. 16-10. This stand has three rings for more stability.

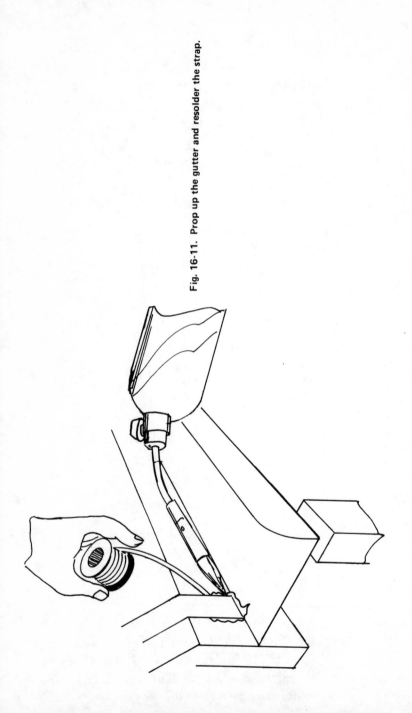

Fig. 16-11. Prop up the gutter and resolder the strap.

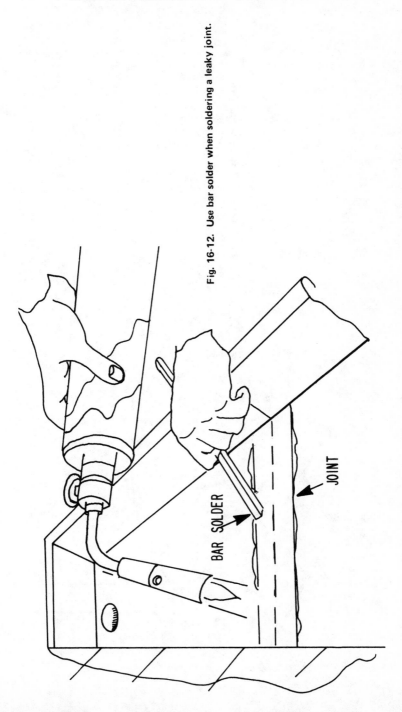

Fig. 16-12. Use bar solder when soldering a leaky joint.

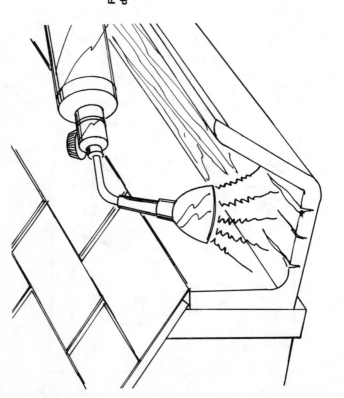

Fig. 16-13. Use a torch flame carefully to dry wooden gutters.

Take proper care of wooden gutters. Cover the exterior with alkyd paint. For the inside of wooden gutters, use black asphalt paint. Apply this paint in hot weather and be sure that the gutter is completely free of moisture. You can use the propane torch to dry the gutter, but use caution (Fig. 16-13). Wood burns readily. Give the gutter a second coat of paint after two days.

Roof leaks can often be attributed to holes in the flashing, which is sheet copper utilized where a roof joins a wall, chimney or adjacent roof. Leaks are often discovered near the bottom section of the valley between roofs. When rain comes down the valley, it carries debris which eventually wears a hole in the flashing. The remedy is to clean the area around the hole, apply flux and use the propane torch to plug the hole with solder. For a big hole, make a sheet copper patch. See Fig. 16-14.

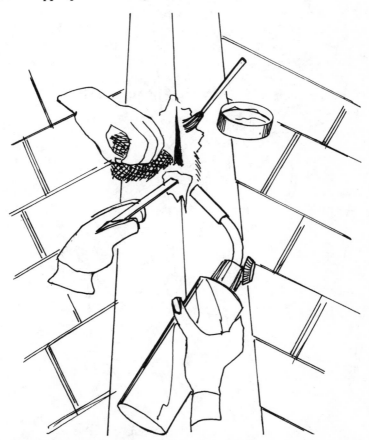

Fig. 16-14. Sheet copper can be used to patch holes in valley flashing.

You can use a technique called *leading* to fix a flashing leak. Leading involves applying a batch of solder over the leaking area (Fig. 16-15). You must sand or wire brush the area until the metal gleams. Then put acid flux over the cleaned surface. Put the tip of a portion of bar solder against the joint. Place the torch's brush flame to the joint, just in front of the bar solder. When a good amount of solder has been melted, move the solder and flame along the seam to leave a patch. Any remaining flux should be rubbed away.

Perhaps you will have to replace defective flashing around vent pipes and chimneys. You can use a propane gas torch to soften the compound. A putty knife does the rest (Fig. 16-16). Put on new flashing compound and take care that all areas have been filled.

Fig. 16-15. Leading is the application of solder over worn-out sections.

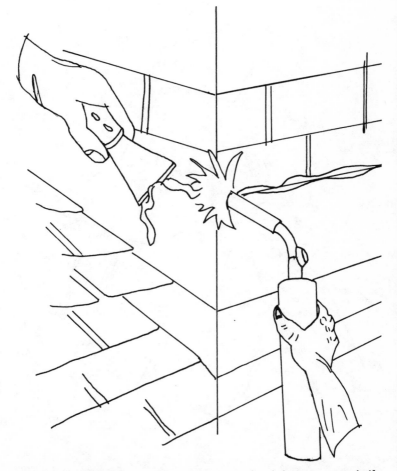

Fig. 16-16. Apply heat to old flashing compound and then use a putty knife to scrape the stuff away.

Another cause of roof leaks is a loose or ripped shingle. You will need to lift the shingle above the worn one in order to make the repair. To prevent cracking, warm the shingle with the torch (Fig. 16-17). Lift the shingle and pry out nails so you can take out the problem shingle. Slide the new shingle into place and nail it. No more leaks!

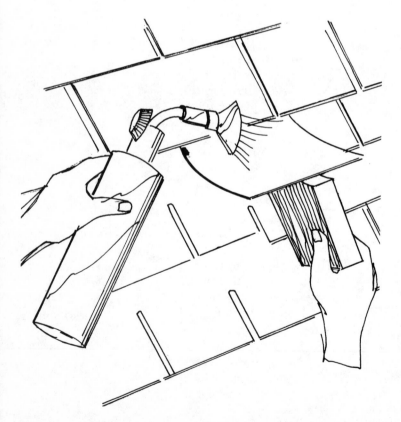

Fig. 16-17. Use the torch to lift the shingle above the damaged one. Heat will prevent the shingle from cracking.

Glossary

abrasion: A condition of wear most often caused by rubbing together of two surfaces.

ac: The accepted abbreviation for *alternating current*. A common type of electricity which reverses its direction of electron flow regularly and periodically.

acetylene: Gas composed of 2 parts carbon and 2 parts hydrogen. When burned in an atmosphere of oxygen, it produces one of the highest flame temperatures available to man.

acetylene cylinder: A specially built container used to store and ship acetylene. Occasionally called "tank" or "bottle."

acetylene regulator: An automatic valve used to reduce acetylene cylinder pressures to working torch pressures and to keep the flow constant.

acid core solder: See preferred term *cored solder*.

activated rosin flux: A rosin or resin base flux which contains an additive to increase wetting action of solder.

alloy: A mixture of two or more metals to achieve specific qualities such as hardness, ductility, etc.

annealing: Softening metals by heat treatment. This process most commonly involves heating the metals up to a critical temperature and then cooling them slowly.

anode: The positive terminal of an electrical circuit.

arc: The flow of electricity through a gaseous space or air gap.

arc cutting: A group of cutting processes wherein the severing or removing of metals is accomplished by melting with the heat from an arc between an electrode and the parent metal.

arc-seam weld: A weld bead with an arc welding unit.

arc-spot weld: A spot weld made by an arc welding process.

arc voltage: The electrical potential across an arc. The pressure or voltage of an arc.

arc welding: Fusing metals using the arc welding process.

automatic oxygen cutting: Oxygen cutting with equipment that is fully automated, requiring only that an operator set up the work initially.

automatic welding: Most commonly some type of arc welding wherein all welding operations are controlled and initiated by automation.

AWS: The abbreviation for the American Welding Society.

axis of a weld: An imaginary line along the center of gravity of the weld metal and perpendicular to a cross section of the weld metal.

back gouging: The forming of a bevel or groove on the other side of a partially welded joint to assure penetration upon subsequent welding from that side.

backfire: Momentary retrogression or burning back of the torch flame into the torch tip. Immediately following the withdrawal of the tip from the work, the gases may be reignited by the hot work piece. Otherwise, the use of a lighter may be necessary.

backhand welding: That method of welding in which the torch and rod are so disposed in the vee that the torch flame points back at the completed weld, enveloping the newly deposited metal. The rod is interposed between the torch and the weld.

balling up: The formation of globules of molten brazing filler material or flux by failure to adequately wet or tin the base metal. Also a professional welders' term (slang) used to describe a poor job.

base metal: Materials composing the pieces to be joined by welding. Also called *parent metal.*

bead: Denotes the appearance of the finished weld and describes the neatness of the ripples formed by the metal while it was in a semi-liquid state.

bevel: A special preparation of metal to be welded wherein the edge is ground or cut to an angle other than 90 degrees to the surface of the parent metal.

blind joint: A joint in which no portion is visible.

blowpipe: A term used to describe an oxyacetylene torch.

bond: Junction of the weld metal and the base metal.

braze: A weld wherein coalescence is produced by heating to suitable temperatures above 800 degrees Fahrenheit and by using a nonferrous filler metal having a melting point below that of the base metals. The filler metal is distributed in the joint by capillary attraction.

braze weld: A weld wherein coalescence is produced by heating to a suitable temperature above 800 degrees Fahrenheit by using a nonferrous filler metal having a melting point below that of the base metals. The filler metal is not distributed in the joint by capillary attraction.

brinell test: A method for determining the surface hardness of metallic materials.

bronze welding: See *braze welding.*

buildup: The amount of weld face or bead which extends above the surface of joined metals.

burned metal: Term occasionally applied to the metal which has been combined with oxygen so that some of the carbon changed into carbon dioxide and some of the iron into iron oxide.

burning: Violent combination of oxygen with any substance which produces heat. This word is sometimes used in place of the term *flame cutting.*

butt joint: An assembly in which the two pieces joined are in the same plane, with the edge of one piece touching the edge of the other.

butt weld: The actual weld in a butt joint.

capillary action: Property of a liquid to move into small spaces if it has the ability to "wet" these surfaces.

calcium carbide: CaC_2. A chemical compound of calcium and carbon usually prepared by fusing lime and coke in an electric furnace. This compound reacts with water to form acetylene gas.

carbon: An element which, when combined with iron, forms various kinds of steel. In steel it is the changing carbon content which changes the physical properties of the steel. Carbon is also used in a solid form as an electrode for arc welding, as a mold to hold weld metal and for brushes in electrical motors.

carbonizing: See *carburizing.*

carburizing: A carburizing flame is an oxygen/fuel gas flame with a slight excess of fuel gas.

case hardening: Adding carbon to the surface of a mild steel object and heat treating to produce a hard surface.

castings: Metallic forms which are produced by pouring molten metal into a shaped container or mold.

cathode: An electrical term for a negative terminal.

celsius: The temperature scale used in the metric system where zero represents the freezing point of water and 100 is the boiling point at sea level. To convert to Fahrenheit, multiply by nine, divide by five and add 32. Celsius was the name of the Swede who invented the centigrade system.

chamfering: See *beveling*.

coalescence: The growing together or growth into one body of the base metal parts.

coated electrode: Metal rod used in arc welding which has a covering of materials to aid in the arc welding process.

complete joint penetration: Joint penetration which extends completely through the joint.

concave fillet weld: A fillet weld having a concave face.

concave root surface: A root surface which is concave.

concave weld face: A weld having the center of its face below the weld edges. An indented weld bead.

cone: The conical part of a gas flame next to the orifice of the tip.

continuous weld: Making the complete weld in one operation.

convex fillet weld: A fillet weld having a convex face.

cored solder: A solder wire or bar containing flux as a core.

corner flange weld: A flange weld with only one member flanged at the location of welding.

corner joint: Junction formed by edges of two pieces of metal touching each other at an angle of about 90 degrees.

corrosive flux: A flux with a residue that chemically attacks the base metal. It may be composed of inorganic salts and acid, organic salts and acids or activated rosins and resins.

coupons: Specimens cut from the weld assembly for testing purposes.

covered electrode: See *coated electrode*.

cracking: The action of opening a valve on a tank of fuel gas or oxygen and then closing the valve immediately.

crater: A depression in the face of a weld, usually at the termination of an arc weld.

creep: The gradual increase of the working pressure (as indicated on the gauge) resulting from the regulator seat not closing tightly against the inlet nozzle and permitting the high pressure gas to leak into the low pressure chamber. When this condition exists, the regulator should be repaired before use.

crown: curve or convex surface of finished weld proper.

cutting attachment: A device which is attached to a gas welding torch to convert it into an oxygen cutting torch.

cutting flame: Cutting by a rapid oxidation process at a high temperature. It is produced by a gas flame accompanied by a jet action which blows the oxides away from the cut.

cutting tip: That part of an oxygen cutting torch from which the gases issue and burn.

cylinder: A portable metallic container for storage and transmission of compressed gases.

dead annealed: The result of heating a work-hardened metal to a red heat and immediately quenching in water. This softens the metal and renders it workable again.

deoxidized copper: Copper from which the oxygen has been removed by the addition of a deoxidizer, phosphorus or silicon. This lowers the electrical conductivity but yields a product more suitable for oxyacetylene welding.

deposited metal: Filler metal that has been added during a welding operation.

depth of fusion: The distance that fusion extends into the base metal or previous pass from the surface melted during welding.

dip brazing: A brazing process in which the heat required is furnished by a molten chemical or metal bath. When a molten chemical bath is used, the bath may act as a flux. When a molten metal bath is used, the bath provides the filler metal.

dip soldering: A soldering process in which the heat required is furnished by a molten metal bath which provides the solder.

direct polarity: Direct current flowing from anode (base metal) to cathode (electrode). The electrode is negative and the base metal is positive.

distortion: Warping of a metal or metal surface as a result of uneven cooling.

downhand welding: Welding in a flat position.

drag: In oxyacetylene cutting, the amount by which the oxygen jet falls behind the perpendicular in passing through the material.

drop-thru: An undesirable sagging or surface irregularity, usually encountered when brazing or welding near the solidus of the base metal. The condition is caused by overheating with rapid diffusion or alloying between the filler metal and the base metal.

ductility: The property of metals that enables them to be mechanically deformed without breaking when cold.

edge joint: A welded joint connecting the edges of two or more parallel or nearly parallel parts.

electrode: A substance which brings electricity up to the point where the arc is to be formed.

elongation: The total amount of stretching of a specimen, produced in a tensile test.

erosion: Reducing size of or wearing away of an object because of liquid or gas impact.

expansion: Increase in one or more of the dimensions of a body, caused usually by a rise in temperature.

explosion welding: A solid state welding process wherein coalescence is effected by high velocity movement produced by a controlled detonation.

face of weld: The exposed surface of a weld.

fahrenheit: A temperature scale used in most English speaking countries where 32 degrees is the temperature at which water will freeze and 212 degrees is the temperature at which it will boil, at sea level. Symbol is F.

ferrous metals: Those metals or alloys of which the principal base or consituent is iron.

filler metal: Material to be added in making a weld.

fillet: Weld metal in the interval vertex, or corner, of the angle formed by two pieces of metal, giving the joint additional strength to withstand unusual stresses.

fillet weld: Metal fused into a corner formed by two pieces of metal whose welded surfaces are approximately 90 degrees to each other.

flame cutting: Cutting performed by an oxygen/fuel gas torch flame which has an oxygen jet.

flanged edge joint: A joint in two pieces of metal formed by flanging the edges at 90 degrees to the plates and joining with an edge weld.

flashback: The retrogression or burning back of the flame into or beyond the mixing chamber, sometimes accompanied by a hissing or squealing sound and the characteristic smoky, sharp pointed flame of small volume. When this occurs, immediately shut off the torch oxygen valve and then the acetylene valve.

flat position: A horizontal weld on the upper side of a horizontal surface.

flowability: The ability of a molten filler metal to flow or spread over a metal surface.

flux: A chemical compound or mixture in powdered, paste or liquid form. Its essential function is to combine with or otherwise render harmless those products of the welding, brazing or soldering operation which would reduce the physical properties of the deposited metal or make the welding, brazing or soldering operation difficult or impossible.

forehand welding: That method of welding in which the torch and rod are so disposed in the vee that the torch flame points ahead in the direction of welding and the rod precedes the torch.

fuel gases: Gases usually used with oxygen for heating such as acetylene, natural gas, propane, methoacetylene, propadyne and other synthetic fuels and hydrocarbons.

fuse plug: A safety device employed on compressed gas cylinders consisting of a low melting point alloy, designed to relieve excessive internal pressure due to heat by melting at a predetermined temperature.

fusion: For the purposes of this book, the melting and flowing together of metals.

gas pocket: A cavity in a weld caused by entrapped fuel gas.

gas welding: A group of welding processes wherein fusion takes place as a direct result of the heat applied with a blowpipe fuel gas and oxygen. It is always best to be specific when discussing welding fuels; therefore oxyacetylene welding, Mapp/oxygen welding, etc.

generator: An apparatus for mechanically controlling the generation of acetylene by the reaction of calcium carbide and water.

gouging: The forming of a bevel or groove by material removed.

groove: The opening provided by a groove weld.

grooved weld: A welding rod fused into a joint which has the base metal removed to form a V, U or J trough at the edge of the metals to be joined.

hard facing or hard surfacing: The application of a hard, wear-resistant alloy to the surface of a softer metal by an arc or gas welding process.

heat: Molecular energy in motion.

heat conductivity: The speed and efficiency of heat energy movement through a substance.

heat-affected zone: That part of the base metal which has been altered by the heat from the welding, brazing or cutting operation but may not have actually melted.

heat conductivity: Speed and efficiency of heat energy movement through a substance.

horizontal position: A weld performed on a horizontal seam at least partially on a vertical surface.

hose: Flexible medium used to carry gases from regulator to the torch. It is made from rubber and reinforced with fabric.

hydrogen: A gas formed of the single element hydrogen. It is considered one of the most active gases. When combined with oxygen, it forms a very clean flame which, however, does not produce very much heat.

icicles: An undesirable condition where excess weld metal protrudes beyond the root of the weld.

incomplete fusion: Fusion which is less than complete.

inclusion: A gas bubble or nonmetallic particle entrapped in the weld metal as a result of improper torch flame or filler material manipulation.

inert gas: A gas which does not normally combine chemically with the base metal or filler metal.

infrared rays: Heat rays which come from both arc and the welding flame.

inside corner weld: Two metals fused together; one metal is held 90 degrees to the other. The fusion is performed inside the vertex of the angle.

intermittent weld: Joining two pieces and leaving unwelded sections in the joint.

joint: Where two pieces meet when a structure is made of smaller pieces.

joint design: The joint geometry together with the required dimensions of the welded joint.

joint penetration: The minimum depth of a groove or flange weld extends from its face into a joint, exclusive of reinforcement.

kerf: The space from which metal has been removed by a cutting process.

keyhole: The term applied to the enlarged root opening which is carried along ahead of the puddle when making an arc weld or other type of welded joint.

knee: The lower arm supporting structure in a resistance-welding machine.

land: The portion of the prepared edge of a part to be joined by a groove weld, which has not been beveled or grooved. Sometimes called *root face*.

lap joint: A welded joint in which two overlapping parts are connected, generally by means of fillet welds.

layer: A certain weld metal thickness made of one or more passes.

lens: A specially treated glass through which a welder may look at an intense flame without being injured by the harmful rays or glare radiating from this flame.

liquidation: The separation of a low melting constituent of an alloy from the remaining constituents, usually apparent in alloys having a wide melting range.

liquefaction: The changing of a gas to a liquid.

liquidus: The lowest temperature at which a metal or an alloy is completely liquid.

low temperature brazing: That group of the brazing processes wherein the brazing alloys employed melt in the range of about 1175–1300 degrees Fahrenheit and the shear type (lap) joint is used.

malleable castings: Cast forms of metal which have been heat-treated to reduce their brittleness.

manifold: A multiple header for connection of individual gas cylinders or torch supply lines.

manual welding: Welding wherein the entire welding operation is performed and controlled by hand.

mapp: A stabilized methyl acetylene-propadiene fuel gas often used in place of acetylene.

melting range: The temperature range between solidus and liquidus.

MIG: A term used to describe gas metal arc welding (metal shielding gas).

mixing chamber: That part of the welding blowpipe where the welding gases are intimately mixed prior to release and combustion.

multilayer welding: In oxyacetylene welding a technique in which a weld, on thick material, is made in two or more passes.

neutral flame: A flame which results from combustion of perfect proportions of oxygen and the welding gas. The most commonly used flame for oxygen/fuel gas welding.

noncorrosive flux: A soldering flux which in itself does not, and with a residue that does not, chemically attack the base metal. It is usually composed of rosin or resin base materials.

nonferrous: Metals containing no substantial amounts of ferrite or iron such as copper, brass, bronze, aluminum and lead.

nozzle: See *tip*.

orifice: Opening through which gases flow. It is usually the final opening or any opening controlled by a valve.

outside corner weld: Fusing two pieces of metal together, with the fusion taking place on the underpart of the seam.

overhead position: A weld made on the underside of the joint with the face of the weld in a horizontal plane.

overlap: Extension of the weld face metal beyond the toe of the weld.

oxidation: The process of oxygen combining with elements to form oxides.

oxide: A chemical compound resulting from the combination of oxygen with other elements.

oxidizing flame: A flame produced by an excess of oxygen in the blowpipe mixture, leaving some free oxygen which tends to burn the molten metal.

oxygen: A gas formed of the element oxygen. When it very actively supports combustion it is called burning; when it slowly combines with a substance it is called oxidation, as in rust.

oxygen-acetylene cutting: Cutting metal using the oxygen jet which is added to an oxygen-acetylene preheating flame.

oxygen-acetylene welding: A method of welding which uses a fuel combination of two gases, oxygen and acetylene.

oxygen cylinder: A specially built container used to store and/or ship oxygen.

oxygen-hydrogen flame: The chemical combining of oxygen with the fuel gas hydrogen.

oxygen hose: See *hose.*

oxygen regulator: An automatic valve used to reduce cylinder pressures to torch pressures and to keep the pressures constant. They are never to be used as acetylene regulators and, in fact, the connections are different.

parent metal: See *base metal.*

pass: Weld metal created by one progression along a weld.

peening: The mechanical working of metal by means of repeated hammer blows.

penetration: The penetration of a weld is the distance from the original surface of the base metal to that point at which fusion ceases.

plug weld: Weld which holds two pieces of metal together. It is made by making a hole in one piece of metal which is then lapped over the other piece.

porosity: Presence of gas pockets or voids in the metal or weld bead.

postheating: Temperature to which a metal is heated after an operation has been performed on the metal such as welding, cutting, forming, etc.

preheating: Temperature to which a metal is heated before an operation is performed on the metal like welding, cutting, forming, etc.

psi: The standard abbreviation for pounds per square inch.

puddle: Portion of a weld that is molten at the place the heat is applied.

quench: To cool hot metal quickly by dunking in a liquid such as water or oil.

reducing flame: An oxygen-fuel gas flame with a slight excess of the fuel gas.

regulator: A mechanical device for accurately controlling the pressure and flow of gases employed in welding, cutting, braze welding, etc.

reinforcement of weld: Weld metal on the face of the weld in excess of that required for the size of the weld, for the purposes of added strength.

resistance welding: A process using the resistance of the metals being welded to the flow of electricity as the source of the heat.

reversed polarity: An electrode positive-anode. Referring to dc and causing electrons to flow from the base metal to the electrode.

root of weld: That part of a weld farthest from the application of weld heat and/or filler metal side.

safety disc: A mechanical safety device designed to release at a predetermined pressure.

skull: The unmelted residue from a liquated filler metal.

slag inclusions: Nonfused, nonmetallic substances in the weld metal.

slugging: The act of adding a separate piece or pieces of material in a joint before or during welding, resulting in a welded joint which does not comply with the original design, drawing or specification requirements.

soldering: A means of fastening metals together by adhering another metal to the two pieces of these metals. The joining metal only is melted during the operation. The joining metal melts below 800 degrees Fahrenheit.

solidus: The highest temperature at which a metal or alloy is completely solid.

spatter: In arc and gas welding, the metal particles expelled during the welding which do not form a part of the weld.

spelter: A term applied to powdered brass used in making a typical brazed joint (lap joint).

spot weld: A weld made between or upon overlapping members wherein fusion may start or occur on the faying surfaces or may have proceeded from the surface of one member. The weld cross section is approximately circular.

straight polarity: An electrode negative-cathode. Connecting dc to cause electrons to flow from the electrode to the base metal.

strain: The reaction of an object to stress.

stress: The load imposed on an object.

stress relieving: Even heating of a structure to a temperature below the critical temperature followed by a slow, even cooling.

surfacing: The deposition of a filler metal on a metal surface to obtain desired properties or dimensions.

sweat soldering: A soldering method in which two or more parts which have been precoated with solder are reheated and assembled into a joint without the use of additional solder.

tack weld: A small weld used to temporarily hold together components of an assembly until they can be welded.

tank: See *cylinder*.

T-joint: A joint formed by placing one metal against another at an angle of 90 degrees. The edge of one metal contacts the surface of the other metal.

tensile strength: Maximum pull stress in psi which a specimen is capable of developing.

throat of fillet weld: Distance from weld face to weld root.

TIG: A term used to describe tungsten inert gas welding.

tinning: In soldering, a coating of the soldering metal given to the metals to be soldered.

tip: Part of the torch at the end where the gas exits and burns, producing the high temperature flame. In resistance welding, the electrode ends are sometimes referred to as the tip.

toe of weld: Junction of the face of the weld and the base metal.

torch: The mechanism which the operator holds during gas welding and cutting, at the end of which the gases are burned to perform the various gas welding and cutting operations. Often called the blowpipe.

ultraviolet rays: Energy waves that emanate from the electrodes and the welding flames of such a frequency that these rays are in the ultraviolet ray light spectrum.

undercut: A depression at the toe of the weld which is below the surface of the base metal.

underfill: A depression on the face of weld or root surface extending below the surface of the adjacent base metal.

vee groove: See *butt joint*.

vertical position: A type of weld where the welding is done on a vertical seam and surface.

welding: The art of fastening metals together by means of interfusing the metals.

weld metal: Fused portion of base metal or fused portion of both the base metal and the filler metal.

weldment: An assembly whose component parts are joined by the welding process.

weld pool: The small body of molten metal created by the flame of the torch.

welding rod: Wire which is melted into the weld metal.

welding sequence: Order in which the component parts of a structure are welded.

work hardening: The increase in strength and hardness produced by working certain metals such as iron, copper, aluminum and nickel. It is most pronounced in cold working.

yield strength: The stress in psi at which a weld specimen assumes a specified limiting permanent set.

Index